INVESTING IN DISASTER RISK REDUCTION FOR RESILIENCE

INVESTING IN DISASTER RISK REDUCTION FOR RESILIENCE
Design, Methods and Knowledge in the face of Climate Change

Edited by

A. NUNO MARTINS
CIAUD (Research Centre for Architecture, Urbanism and Design), Faculty of Architecture, University of Lisbon, Lisbon, Portugal

GONZALO LIZARRALDE
École d'Architecture, Université de Montréal, Montreal, Quebec, Canada

TEMITOPE EGBELAKIN
School of Architecture and Built Environment, University of Newcastle, Newcastle, NSW, Australia

LILIANE HOBEICA
RISKam (Research Group on Environmental Hazard and Risk Assessment and Management), Centre for Geographical Studies, University of Lisbon, Lisbon, Portugal

JOSÉ MANUEL MENDES
Centre for Social Studies and Faculty of Economics, University of Coimbra, Coimbra, Portugal

ADIB HOBEICA
Independent Consultant, Coimbra, Portugal

ELSEVIER

Elsevier
Radarweg 29, PO Box 211, 1000 AE Amsterdam, Netherlands
The Boulevard, Langford Lane, Kidlington, Oxford OX5 1GB, United Kingdom
50 Hampshire Street, 5th Floor, Cambridge, MA 02139, United States

Notices

Knowledge and best practice in this field are constantly changing. As new research and experience broaden our understanding, changes in research methods, professional practices, or medical treatment may become necessary.

Practitioners and researchers must always rely on their own experience and knowledge in evaluating and using any information, methods, compounds, or experiments described herein. In using such information or methods they should be mindful of their own safety and the safety of others, including parties for whom they have a professional responsibility.

To the fullest extent of the law, neither the Publisher nor the authors, contributors, or editors, assume any liability for any injury and/or damage to persons or property as a matter of products liability, negligence or otherwise, or from any use or operation of any methods, products, instructions, or ideas contained in the material herein.

Library of Congress Cataloging-in-Publication Data
A catalog record for this book is available from the Library of Congress

British Library Cataloguing-in-Publication Data
A catalogue record for this book is available from the British Library

ISBN: 978-0-12-818639-8

For information on all Elsevier publications visit our website at
https://www.elsevier.com/books-and-journals

Publisher: Candice Janco
Acquisitions Editor: Candice Janco
Editorial Project Manager: Isabella C. Silva
Production Project Manager: Surya Narayanan Jayachandran
Cover Designer: Christian J. Bilbow

Typeset by TNQ Technologies

Contents

Contributors

Olalekan Adekola
Department of Geography, York St John University, York, United Kingdom

Maria do Céu Almeida
National Laboratory for Civil Engineering (LNEC), Lisbon, Portugal

Dilanthi Amaratunga
Global Disaster Resilience Centre, University of Huddersfield, Huddersfield, United Kingdom

Neide Portela Areia
Centre for Social Studies, University of Coimbra, Coimbra, Portugal

Rita Salgado Brito
National Laboratory for Civil Engineering (LNEC), Lisbon, Portugal

Stephen Buckman
Clemson University, Greenville, South Carolina, United States

Yus Budiyono
Agency for the Assessment and Application of Technology, Jakarta, Indonesia

Maria Adriana Cardoso
National Laboratory for Civil Engineering (LNEC), Lisbon, Portugal

Sandra Carrasco
School of Architecture and Built Environment, University of Newcastle, Newcastle, NSW, Australia

Temitope Egbelakin
School of Architecture and Built Environment, University of Newcastle, Newcastle, NSW, Australia

Matthias Garschagen
Department of Geography, Ludwig-Maximilians University of Munich, Munich, Germany

Momoyo Gota
Tokyo University of Science, Tokyo, Japan

Charlotte Kendra Gotangco
Department of Environmental Science, School of Science and Engineering, and Ateneo Institute of Sustainability, Ateneo de Manila University, Quezon City, Metro Manila, Philippines

Adib Hobeica
Independent Consultant, Coimbra, Portugal

Liliane Hobeica
RISKam (Research Group on Environmental Hazard and Risk Assessment and Management), Centre for Geographical Studies, University of Lisbon, Lisbon, Portugal

Jairus Carmela Josol
Department of Environmental Science, School of Science and Engineering, and Ateneo Institute of Sustainability, Ateneo de Manila University, Quezon City, Metro Manila, Philippines

Tomoko Kano
Teikyo Heisei University, Tokyo, Japan

Kaushal Keraminiyage
University of Salford, Salford, United Kingdom

Jessica Lamond
Department of Architecture and the Built Environment, University of the West of England, Bristol, United Kingdom

Allan Lavell
FLACSO, San José, Costa Rica

Gonzalo Lizarralde
École d'Architecture, Université de Montréal, Montreal, Quebec, Canada

Arlene Christy D. Lusterio
TAO-Pilipinas (Technical Assistance Organization), Inc., Quezon City, Metro Manila, Philippines

A. Nuno Martins
CIAUD (Research Centre for Architecture, Urbanism and Design), Faculty of Architecture, University of Lisbon, Lisbon, Portugal

Geraldine R. Matabang
TAO-Pilipinas (Technical Assistance Organization), Inc., Quezon City, Metro Manila, Philippines

José Manuel Mendes
Centre for Social Studies and Faculty of Economics, University of Coimbra, Coimbra, Portugal

Marco Morais
Municipality of Lisbon (CML), Lisbon, Portugal

Olabode Ogunmakinde
Faculty of Society and Design, Bond University, Robina, QLD, Australia

Hugo Pinto
Centre for Social Studies and Faculty of Economics, University of Coimbra, Coimbra, Portugal

Pournima Sridarran
University of Moratuwa, Moratuwa, Sri Lanka

Gusti Ayu Ketut Surtiari
Research Center for Population, Indonesian Institute of Sciences, Jakarta, Indonesia; University of Bonn, Bonn, Germany

Takae Tanaka
Tama University, Kanagawa, Japan

Ma. Theresa Amor J. Tan Singco
Disaster Risk Reduction Network Philippines (DRRNetPhils), Quezon City, Metro Manila, Philippines

Alexandre Oliveira Tavares
Centre for Social Studies and Faculty of Sciences and Technology, University of Coimbra, Coimbra, Portugal

Maria João Telhado
Municipality of Lisbon (CML), Lisbon, Portugal

Foreword

Allan Lavell
FLACSO, San José, Costa Rica

Study and practice on disasters and disaster risks have changed in emphasis dramatically over the past 60 years. From a dominant concern for **disaster**, preparedness, response, and physical reconstruction, increasing emphasis has been placed on disaster **risk**, the latent preamble to disasters. This includes the consideration of the processes by which risks are constructed or created, notably the study of underlying structural causes and derived risk drivers, and the types of intervention required for their control (prevention) or, where already existing, reduction (mitigation). This also includes proposals for planned post-disaster recovery, in particular physical reconstruction, which seek to avoid future risks and achieve such goals as "building back better" and "transformational recovery." The present COVID-19 crisis, marked by strong social and economic impacts at the regional, national, and global scales, has led to a serious ongoing debate on these needs and emphases.

Such a transition has been possible thanks to the development of concepts on vulnerability and socially conditioned exposure to hazard, and the idea of socionatural hazards. Taken together and incorporating notions of the social perception of hazard and risk, these theoretical foundations constitute what is known as a social-construction perspective for understanding and intervening on risks and disasters. Such a perspective opens up the option for potentially controlling risk construction and deconstructing risk where it already exists. However, this option is circumscribed by the need for culturally, socially, economically, politically, and historically sensitive understandings of risk. It also depends on the circumstances in which different social, ethnic, racial, and other groups live and exist, their needs and priorities, and the intersectional relations between different human conditions related to such aspects as gender, age, and disability.

Moreover, from the outset, it should be emphasized that concepts and discourses on disaster risk and its prevention or reduction and transformative recovery are far ahead of real practice. Numerous studies and databases show that risks and disasters increase more than arithmetically as time passes by. It is no surprise that this is so, as everyday, chronic, or quotidian risk—expressed in such conditions as poverty, addiction, social and personal violence, and ill health—closely correlate with extensive and intensive disaster risk. These are further synergized through prevailing and growing conditions of inequality in access to voice and power, recognition, basic services, income, and employment. But overcoming these conditions is an imperative for sustained risk reduction and new risk-controlled development.

International agreements or frameworks on disasters and disaster risks—from UNDRO's early works in the 1970s and 1980s, through the International Decade and Yokohama in the 1990s, to the International Strategy and the Hyogo and Sendai accords, between 2000 and the present—have taken up on and reflected in their conceptual frames many of the new ideas that have been generated in academia and practice over the years. For instance, Hyogo made the reduction of "underlying risk factors" one of its five central objectives and laid out in its framework many of the critical areas and processes that generate and construct risks in society (UNISDR, 2005). Sendai forsook such a direct reference to "underlying factors" and, for the reasons that be, prioritizes increased "investment in disaster risk reduction for resilience" (UNISDR, 2015).

At the same time, as agreement passed to agreement over time, the notion of "natural" disaster was left behind (with Hyogo), at least in headlines and titles, and new or derived concepts such as resilience (Hyogo) and systemic risk (Sendai) appeared in the discourses, many times uncritically and poorly defined in the beginnings. What rarely appears or is partially dealt with in UN and other international agreements is the direct and critical reference to the underlying structural conditions that define many so-called development models. Such conditions are at the root of much disaster risk and explain why action to reduce or control risk often fails or is not even considered in planning formats.

This is of course understandable as the raising of such issues would be tantamount to a criticism of self, even if couched in historical terms. On the other hand, it would lead to the imperative for many stakeholders, at very different levels and scales, to accept that risk is endogenous to development or economic-growth processes and not an exogenous creation relating to external hazards impinging on innocent societies—see Hewitt (1983). This idea should make it very clear to us that disaster risk management (DRM), particularly when considering prospective risk avoidance, is basically a false or "exotic" notion given that the real problem is the type and direction of development. Risks and disasters are actualizations, materializations, manifestations, reflections, symbols, and representations of such development malaise and skewed development processes. This means that as regards disaster risk reduction and avoidance—as opposed to preparedness, response, and even aspects of recovery—traditional disaster professionals and processes cannot hope to tackle a problem that basically relates to the need for sustainability in development and new bases for its promotion. The COVID-19 pandemic again has made this very clear. The increasing demand for green recovery, humanitarian and climate-adapted design, and other modalities of more just recovery are reflections of needs historically formulated but brought to the forefront due to the present crisis.

The former arguments and conclusions, among others, set the scene for a consideration of the pertinence, relevance, gaps, and opportunities that international agreements and their proposed actions, in this case the Sendai Framework and its Priority 3, offer in terms of real advance toward DRM goals. And here a series of aspects must be put on the table as additional background to the reading and analysis of the chapters of the present book.

For instance, international agreements seek to achieve consensus at a government level on the main factors in play, the main goals to be achieved, and the main routes to be taken to overcome problems and stumbling blocks. Yet, although being informed by wider discussion involving diverse stakeholders from the private sector and civil society, these agreements are always limited or constrained by at least three major circumstances.

Firstly, they are global and do not consider local, regional, or national circumstances that are fundamental to take into account in searching for the route to successful action in differing circumstances. Second, they are generic as opposed to specific and normally are not prescriptive but rather suggestive. And thirdly, they do not self-criticize and, therefore, by nature ignore or sideline structural conditions that hinder advance or create negative circumstances.

The challenge for those working in the disaster-risk area—whether academics, practitioners from NGOs or other institutions, or government agents at local and national levels—is to interpret, filter, and specify generic ideas and pathways taking into consideration different territorial and social scales and circumstances. This must be done accepting that disaster risk is not an autonomous state of affairs with its own, independent drivers. Rather it is a derived context, in which structural conditions and everyday needs and demands condition how risks are understood, constructed, and how to deal with them and with what methods. And many of these solutions are not directly DRM-related or guided but, rather, reflect the need to achieve—through sector and territorial, social, economic, and environmental planning and using coproduced, consensual methods—livelihood opportunities and satisfiers, even if at the cost of loss and damage associated with hazards in some longer-term future.

This is and will be the situation until we achieve a world where severe inequality, unsustainability, and injustice do not reign so prevalently as they do today in too many parts. The interventions we make today in favor of risk reduction and control, guided by the principles of increased equality and access, human rights, and justice, will help guide the path to such a future condition. The chapters in this book should serve to promote that end. The book's three parts—focused on architectural and urban design, methods, and knowledge—timely provide evidence of how to transform words into practice and scale down notions, guidelines, and priorities established at higher levels of resolution and social discourse.

References

Hewitt, K. (1983). *Interpretations of calamity from the viewpoint of human ecology.* Boston, MA: Allen & Urwin.

UNISDR (United Nations Office for Disaster Risk Reduction). (2005). *Hyogo Framework for Action 2005–2015: Building the resilience of nations and communities to disasters.* Retrieved from https://www.unisdr.org/files/1037_hyogoframeworkforactionenglish.pdf.

UNISDR (United Nations Office for Disaster Risk Reduction). (2015). *Sendai Framework for Disaster Risk Reduction 2015–2030.* Retrieved from https://www.preventionweb.net/files/43291_sendaiframeworkfordrren.pdf.

Preface

Public and private investments in disaster risk prevention and reduction through structural and non-structural measures are essential to enhance the economic, social, health and cultural resilience of persons, communities, countries and their assets, as well as the environment. These can be drivers of innovation, growth and job creation. Such measures are cost-effective and instrumental to save lives, prevent and reduce losses and ensure effective recovery and rehabilitation.

Summary of the Sendai Framework's third priority (UNISDR, 2015).

1. Where good intentions meet reality

A few years ago, we gathered a group of about 30 experienced researchers—this book's chapter contributors—to explore the challenges and opportunities that emerge in the implementation of the Sendai Framework for Disaster Risk Reduction. More than three years had passed since the Sendai Framework was launched with much fanfare and optimism, and we had pressing questions at hand: Is it actually working? Are politicians truly interested in investing in risk reduction? How are investments in risk reduction competing with postdisaster response (which often yields more political value)?

At that time, four of the six editors of this book had chaired a global scientific meeting structured around the four priorities of the Sendai Framework: the 8th International Conference on Building Resilience (Lisbon, Portugal, November 2018). The event contributed to the publication of a collection of four books, each named after one of the Sendai priorities. Our team focused on the Sendai Framework's third priority, which provides the title for this book. We quickly realized that, of course, most people, agencies, and organizations want to prevent disasters and avoid suffering. Yet, by the way it was written, the Sendai Framework prompted additional questions: What do we do when we decide to "invest" in disaster risk reduction (DRR)? Which political, economic, and ethical dimensions are revealed by our intention of "investing" in "resilience?" What gets masked (or underrated) by the narratives of "public and private investments," "innovation, growth, and job creation?"

These questions guided our work for the past few years. Across 11 chapters, we can now provide examples of projects, programs, and policies that have channeled money and efforts to reduce risks and enhance resilience in different parts of the world—from the Philippines to Portugal, and from Haiti to New Zealand.

So we do have some answers and recommendations. But not of the simple kind. Through these cases, you will discover how difficult it is to invest in DRR and how contentious it is to look for resilience. The road to disaster resilience is paved as much with good intentions as with problematic implementation. The challenges at hand posed by the prioritization of investments in postdisaster response over prevention and mitigation are illustrated by the ongoing COVID-19 pandemic, which absorbed countries' humans and financial resources at a level we had not seen for a long time. Indeed, the concentration of funds in the fight against the pandemic caused significant cuts in international aid and assistance programs (ranging from 30% to 80%), thereby affecting the work of many NGOs and international agencies largely dependent on developed countries' donations.

The Sendai Framework and other policy documents drafted at the global level can be useful to set up a common language, identify patterns, and inspire others toward change. But policy is not enough—even when it is sealed with a United Nations logo. Disaster risk reduction and resilience must be achieved by human beings, not reports and frameworks. And this is where good intentions face challenges. Government officers, professionals, NGO representatives, local leaders, politicians, and citizens are, of course, heterogeneous and often unpredictable. Their interpretations of risk, danger, disaster, resilience, and sustainability are diverse, dynamic, and linked to personal interests, needs, and expectations. This book is about the space where good intentions (consecrated in international policies) meet on-the-ground human realities.

Unlike what many officers in international agencies and the authors of the Sendai Framework would like us to believe, the road toward DRR is neither straight nor unique. In this book, we will see that there are as many ways of responding to danger and hazards as there are stakeholders and places where they operate.

Every context is, of course, unique. But it is still possible to identify some patterns. In fact, we edited this book to explore those common traits and learn from them. Here is a pattern that helped us structure this book: The cases we present attempt to reduce risk or rebuild after disasters in three distinctive ways.

First, by exploring design solutions, notably participatory processes, opportunities, and tensions that emerge in architectural and urban projects that deal with disaster risk (often known as humanitarian design), as well as in the search for healthier relationships between culture, ecosystems, and the built environment. Second, by devising and implementing new methods and innovative practices to address DRR. Finally, by investing in knowledge, information, and communication to create better policy and governance mechanisms. Consequently, the chapters in this book are structured around three parts. In each part, the authors uncover real-life experiences in which conflicting objectives, fragmentation, lack of common goals, and other implementation barriers occurred. You will learn how, in some cases, stakeholders managed to overcome them. In other cases, the road was paved with good intentions but little positive impact.

2. Transformation through innovative design and planning is sometimes needed, but so are heritage and continuity

Good architectural and urban design is required to reduce risks and rebuild after a disaster. But what is a good design or plan when it comes to avoiding damage or reconstructing after a disaster? In the first part of the book, we explore the role of design in DRR and pay attention to how it can help protect heritage areas and create the conditions for maintaining architectural and urban cultural values.

Chapter 1 explores the objective of investing in community participation for disaster recovery. Lusterio, Matabang, and Tan Singco focus on how disaster recovery through participatory planning and design can support DRR decisions and investments. They observe that changes in DRR policy in the Philippines might threaten grassroots efforts, rendering difficult the implementation of the Sendai Framework's principle of "all-of-society engagement and partnership" (UNISDR, 2015). They find advantages, but also drawbacks, in decentralized DRR structures, including duplications of efforts and difficulties in prioritizing actions. Yet they also find that stakeholders can mitigate these drawbacks by adopting a "consortium approach" involving citizens and valuing the communities' voices and decision-making power.

Chapter 2, by Kano, Tanaka, and Gota, sheds light on the potential of social places in heritage cities to better integrate risk management. The authors address the difficulties of providing safe refuge in historic cities. The lack of possibilities for new infrastructure projects led them to consider the potential of open and social spaces in urban settings to promote safe and resilient refuges in the event of a disaster. Therefore, this study integrates residents, businesses, and tourists into DRR efforts as active actors in the enhancement of community resilience and the promotion of sustainable heritage tourism.

Chapter 3 examines a landmark flood-protection intervention in a historic site in Portugal. By doing so, Hobeica and Hobeica question shortsighted stances on risks and heritage sites. They posit that the coexistence between built assets and flood hazards is a legacy that is worth investing to maintain. Challenging the idea of preventing disasters at any cost, the authors claim that embracing flood risk when designing for the preservation of historical assets can be more beneficial. This can strengthen cultural resilience. Investments that incorporate contingencies instead of avoiding them help raise awareness about vulnerabilities and DRR while preserving historic sites and allowing visitors to enjoy them during (unavoidable) flood events.

3. Methods depend on meanings, perspectives, and agendas

Mitigation, adaptation, prevention, avoidance, acceptance... There are many attitudes that we can exhibit toward natural hazards and climate-change effects. But stakeholders do not necessarily agree on how to adopt these reactions or the ways of explaining and justifying them. These differences render implementation difficult and cause tensions. In the second part of the book, we explore how different methods to understand and act upon risks can overcome fragmentation, red tape, conflicts, tensions, and other common implementation barriers.

In Chapter 4, in line with the Sendai Framework's guiding principle of "a multihazard approach" in DRR (UNISDR, 2015), Gotangco and Josol explore the cascading effects of multiple flooding hazards and their relationships with urban services through the use of a Physical Services Index. Taking Metro Manila (the Philippines) as an example, the authors argue that relying on DRR infrastructures is insufficient as a flood-mitigation strategy, due to demographic changes, urbanization trends, and other pressures that affect these solutions in the medium and long terms. Echoing conclusions presented in Chapter 1 by Lusterio and colleagues, the authors contend that DRR in Metro Manila should be oriented toward vulnerability reduction, rather than simple hazard mitigation.

Chapter 5 explores tensions between people's objectives and stresses the critical role of communication in DRR and climate change adaptation (CCA). Buckman finds that technical jargon used in CCA often hinders positive change in cases in which communities do not have the same goals. He warns that pursuing the Sendai Framework's guiding principle of coherence in the "development, strengthening and implementation of relevant policies, plans, practices and mechanisms" (UNISDR, 2015) can be problematic, notably when terms about climate change fail to reflect local meanings, culturally accepted concepts, and religious ideas. A common and locally agreed-upon terminology is sometimes required to align objectives and work toward common goals. Scenario planning and inclusive design practices help bringing together dissenting voices.

Sharing methodological similarities with Chapters 4 and 5, Chapter 6, explores systems thinking as a means to improve DRR and CCA practices and communication. Adekola and Lamond argue that differences in perceptions and understanding of risk and climate change pose a real challenge for building resilience. It is therefore necessary to "move beyond the usual debate around cause, effect, and solution, to encompass exploring future scenarios and communication." They point out that by embracing the "empowerment of local authorities and communities" among its guiding principles, the Sendai Framework implicitly condemns existing power imbalances (UNISDR, 2015). But it does not delve into the sociopolitical roots of such imbalances, a fact that constitutes a barrier for its implementation.

In Chapter 7, Cardoso, Almeida, Telhado, Morais, and Brito analyze the impacts of climate change on cities, specifically on urban water cycles. Based on the results of the RESCCUE research project, they apply the Resilience Assessment Framework to Lisbon (Portugal), in the mobility and waste sectors. They explore functional and physical dimensions of resilience concerning the impacts of extreme hydrological events. The authors conclude that the Resilience Assessment Framework is a key tool to allocate resources to the design and implementation of DRR strategies, plans, and policies.

4. Who knows what?

Knowledge is a source of power. DRR and CCA responses largely depend on who controls knowledge, information, and communication. Power imbalances and differentiated access to expertise and information often frame the governance conditions in which DRR interventions are designed. As illustrated in the third part of the book, these imbalances play a role in how investments are made (see Chapter 8), how people react to relocation programs (see Chapter 9), how people understand climate change (see Chapter 10), and who gets to participate in transformative action (see Chapters 9 and 11).

In Chapter 8, Egbelakin, Ogunmakinde, and Carrasco look into the awareness and preferences of property owners regarding existing incentives for the seismic retrofitting of heritage buildings in New Zealand. They highlight the complexity of DRR investments, as these are shaped by governance mechanisms, risk perception, and people's willingness to commit financial resources. Like Chapter 5, this chapter also emphasizes how communication—both message and means—is essentially a political act and plays a crucial role in risk reduction. Yet, this aspect is underrated in the Sendai Framework. By avoiding to touch upon the political component of DRR, the Sendai Framework can perpetuate the status quo and its many sources of "disaster risk creation"—for more on this argument, see also Wisner and Lavell (2017).

Chapter 9 reveals failures in, and people's dissatisfaction with, postdisaster resettlements. Building on several Sri Lankan cases, Sridarran, Keraminiyage, and Amaratunga highlight some of the shortcomings of conflict and postdisaster displacements and resettlements. The authors show that although resettlers' expectations in the early stages of the process are often high, they tend to fade away when the mismatches between actual needs and the overall conditions of the new built environment become tangible. During the displacement and transition periods, the new settlers often overrate some favorable conditions, such as being entitled to own a house for free. Yet, they later realize the consequences of unfavorable conditions, such as distance from jobs or hazard exposure. This chapter illustrates how the Sendai Framework's guiding principle of "building back better" in postdisaster recovery can become an empty slogan that is hard to materialize (UNISDR, 2015).

In Chapter 10, Tavares, Areia, Mendes, and Pinto analyze the role of the media in Portugal in disseminating climate-change information and promoting public awareness and engagement.

Individuals can only be active CCA players if they properly understand the issues at stake. But the authors' study shows that policy-makers' interests often prevail over scientific (evidence-based) climate-related information. This bias in communication hinders public awareness, potentially hampering both individuals' shift toward more sustainable lifestyles and collective participation in climate governance.

Finally, in Chapter 11, Surtiari, Garschagen, Mendes, and Budiyono deal with the evaluation of governmental CCA interventions. Considering adaptation as a long-term endeavor whose outcomes are hard to predict, the authors call for reviewing interventions by adopting an inclusive approach that considers its impacts on different populations and timeframes. Through the lens of vulnerability, they explore the impact of adaptation measures on people living in slums in coastal Jakarta (Indonesia). They reveal changes in slum dwellers' risk perception and informal institutions, after the implementation of formal hard-engineering adaptation measures. They show that people's vulnerability tends to increase when adaptation programs underrate the role played by informality. The likelihood of maladaptation is high when investments focus on hazard mitigation and neglect the needs, practices, and capacities of those most vulnerable to the impacts of climate change.

5. There is no easy way out

When it comes to investing in DRR, there is no magic bullet. Our call is to fully embrace the complexity and dynamic character of DRR and climate-change action. We must be wary of frameworks that lead us to believe that consensus on action already exists and that implementation is only a subproduct of policy, simply waiting to happen. The cases presented in this book demonstrate that consensus rarely exists regarding goals and ways to achieve them. And implementation is not a matter of putting together a few pieces after a policy is written. It requires proper governance mechanisms, sustained dialogue, transparency, innovation, commitment, resources, and time and persistence... lots of them. But above anything else, DRR requires an ethical and political stance that recognizes the social and environmental injustices that create risks. Welcome to the challenges of investing in disaster risk reduction.

The editors, 2022

References

UNISDR (United Nations Office for Disaster Risk Reduction). (2015). *Sendai Framework for Disaster Risk Reduction 2015–2030*. Retrieved from https://www.preventionweb.net/files/43291_sendaiframeworkfordrren.pdf.

Wisner, B., & Lavell, A. (2017). *The next paradigm shift: From 'disaster risk reduction' to 'resisting disaster risk creation' (DRR > RDRC)*. Keynote speech presented in the Dealing with Disasters Conference (University of Durham, UK, September 2017). Retrieved from https://www.researchgate.net/publication/320045120_The_Next_Paradigm_Shift_From_'Disaster_Risk_Reduction'_to_'Resisting_Disaster_Risk_Creation.

Acknowledgments

We dedicate this book to the community of Gamboa de Baixo, in Salvador da Bahia, Brazil, depicted on the cover page (photograph by A. Nuno Martins). This informal area, inhabited for centuries by fisherfolk, was built over the foundations of an old Portuguese and Dutch fortress. It illustrates and somehow synthesizes the challenges faced by vulnerable communities highly exposed to various hazards. This book does not address the achievements of Gamboa's householders in disaster resilience and heritage conservation. Nevertheless, Gamboa de Baixo represents all those slums, neglected central areas, and heritage sites focused on in the following chapters and many similar disaster-prone spots across the Global South. To all of them, we humbly offer these research works and express our deep respect. Their humanitarian and architectural accomplishments related to disaster risk reduction feed our scientific minds and inspire us to build connections among people, knowledge, and resources.

The editors also acknowledge the NGO Building 4Humanity Designing and Reconstructing Communities Association, the University of Lisbon, the University of Coimbra, the University of Huddersfield, and the Portuguese Foundation for Science and Technology, for making possible the 8th International Conference on Building Resilience in Lisbon (November 2018).

The last acknowledgments go to our colleagues Pedro Pinto Santos and Jo Rose for their initial involvement and contributions to this book.

Introduction

Investing in disaster-risk consultants and visibility

Gonzalo Lizarralde
École d'Architecture, Université de Montréal, Montreal, Quebec, Canada

1. More than slogans: the root causes of disasters

"The overall conclusion is that both the [Hyogo Framework for Action] and the [Sendai Framework for Disaster Risk Reduction] fail to deal with root causes of disaster."

These are the words used in 2020 by Ben Wisner, a professor at University College London, after evaluating five years of implementation of the Sendai Framework (Wisner, 2020, p. 239). Then he asked: "Should we not take a critical step back from mottos, oft-repeated phrases, neologisms, and metaphors that may serve as prison bars, walls, and jailors?" "Can we get beyond frameworks?" he finally wondered (Wisner, 2020, p. 240).

This book argues that Wisner is right about the Sendai Framework. And the third Sendai priority, "investing in disaster risk reduction for resilience," is no exception (UNISDR, 2015, p. 14).

The priority that provides the title for this book is full of the empty jargon that characterizes UN-sanctioned international policy frameworks. This priority recognizes that "structural and non-structural measures are essential to enhance the economic, social, health and cultural resilience of persons, communities, countries and their assets, as well as the environment" (UNISDR, 2015, p. 18). There are references to the importance of preserving historic places and cultural heritage, and eradicating hunger and poverty. But perhaps more importantly for the proponents of the framework itself, all the keywords to please economists, investors, and politicians are also there: innovation, economic growth, risk transfer, risk sharing, financial protection, public and private investment, and, of course, business resilience. And there is plenty for those interested in globalization and universalized ideas: It is

Investing in Disaster Risk Reduction for Resilience. https://doi.org/10.1016/B978-0-12-818639-8.02001-9

necessary to "encourage the revision of existing or the development of new building codes and standards"; and also "the integration of disaster risk reduction considerations and measures in financial and fiscal instruments" (UNISDR, 2015, pp. 19–20). How to do all of this? Apparently, resilience will do the trick, but the framework falls short in explaining why or how. Another part of the answer lies on "close cooperation with partners in the international community, business, international financial institutions and other relevant stakeholders" (UNISDR, 2015, p. 20).

Concerning disaster response, the Sendai Framework also has a convenient prescription: "In the post-disaster recovery, rehabilitation and reconstruction phase, it is critical to prevent the creation of and to reduce disaster risk by 'Building Back Better'" (UNISDR, 2015, p. 14).

Pertinent ideas, you might think. But are they? Have they really worked?

Since 2015, when the Sendai Framework was elevated to international policy status by the United Nations, things have gotten worse. That same year, an earthquake killed more than 8000 people in Nepal. The year after, another earthquake destroyed several places in Ecuador. In 2017, Hurricane Maria caused an undetermined number of deaths in Puerto Rico. Wildfires hit hard several Greek locations in 2018. The same year, an earthquake ruined the region of Lombok in Indonesia and droughts in Australia killed millions of animals. Fires in the Amazon Region and Australia destroyed millions of hectares of forest, killing an unknown number of animals and affecting endangered species. In 2019, Hurricane Dorian devastated the northern Bahamas Islands and some parts of the United States. That same year, Tropical Storm Imelda ravaged Texas, and California experienced mighty wildfires. In 2020, Hurricane Iota brought massive destruction in the Caribbean. In 2020 and 2021, the COVID-19 pandemic caused deaths, illness, and economic paralysis almost worldwide. In 2021, a winter storm led to blackouts and several other disruptions in Texas and other areas of the United States.

To be sure, these events killed less people than similar events in the past. But the amount of destruction and the economic impacts did increase, particularly in developed countries (Hannah & Roser, 2019). Between 2015 and 2019, the United States "experienced 69 separate billion-dollar disaster events" (Smith, 2020). The five-year mean cost of disasters in the United States almost doubled during that period. The average number of fatalities per year has also increased, from 217 in the 1990–1999 period to 772 in the 2015–2019 period—with increases in every single period in between (Smith, 2020). Several small disasters also ravaged cities and villages everywhere.

It might be unfair to use these disasters as a proof of the ineffectiveness of the Sendai Framework. In fact, the Sendai Framework was still in its infancy when many of these disturbing events happened. It was perhaps too soon to see transformations on the ground to prevent all this suffering and destruction. It is perhaps fairer to assess the Sendai Framework and the international agreement on "investing in disaster risk resilience" based on the results of actions following a major disaster that happened a few years before 2015, mobilized almost all UN institutions, international cooperation agencies, and international banks, attracted significant global attention, and was at a certain moment seen as an example of Building Back Better. Such a case would allow us to see how different stakeholders have invested in disaster risk reduction for resilience. It would show us how international agreements on risk reduction can operate to prevent future casualties and destruction. Well, as it happens, such an event does exist. Locals called it Goudougoudou (a Haitian Creole onomatopoeia) and it occurred in what some observers dubbed "the Republic of NGOs" (Thomas, 2020).

2. The mighty Goudougoudou

The January 2010 disaster that largely destroyed Port-au-Prince was exceptional in many ways. It killed an undetermined number of people (estimates vary between 60,000 and 200,000), destroyed the national palace, paralyzed a whole country, and brought the Haitian government to a stall. Unfortunately, it followed a pattern of destruction after natural hazards that has become all too frequent in Haiti: Hurricane Hanna had caused massive destruction two years earlier and so did Hurricane Mathew in 2016. Landslides and floods are frequent after heavy rains, whereas some parts of the country are often hit by long droughts. After the 2010 earthquake, a cholera epidemic followed, and, of course, Haiti was also hit by the COVID-19 pandemic.

The earthquake brought significant attention worldwide. Even before 2010, Haiti was flooded with charities; but soon after the earthquake, hundreds of additional NGOs jumped in. Physicians and other medical workers volunteered to work during the emergency phase. Virtually all the main international agencies set up teams and local offices in Port-au-Prince. Millions of dollars in donations were promised in the United States, Canada, and

Europe. Foreign private companies offered houses, latrines, solar panels, construction materials, and other "solutions" to the housing crisis. Urban planners, architects, and engineers volunteered to design infrastructures, housing, hospitals, and schools. Construction associations and professional bodies offered help to upgrade building codes and regulations. Bill Clinton, Sean Penn, and other celebrities got personally involved in both raising money internationally and showing support on the ground. Numerous religious organizations (mostly Christian) collected funds to build schools, kindergartens, and temples. The World Bank, the International Monetary Fund, and other international agencies conducted studies, guidelines, pathways, and reports on how to help the country. The Clinton Foundation, the Prince Charles Foundation, and other well-known charities set up a priority to help Haitians "build back better." For a moment, it looked like all the ingredients were available to invest in disaster risk reduction for resilience.

When the Sendai Framework was officially launched in 2015, the reconstruction process in Haiti was still incomplete. There was a clear opportunity for international agencies and banks to demonstrate the merits of investing in disaster risk reduction for resilience. By then, my research team at Université de Montréal and I had been working in Haiti for more than five years. But what we witnessed in 2015 was exactly the opposite of an interest in investing in disaster risk reduction. In fact, five years after the earthquake, most NGOs, international agencies, and expatriates who had arrived in Port-au-Prince were leaving the country, closing activities, and laying down local staff. Many were packing to set up activities in Nepal. In fact, Hurricane Sandy had already shifted the focus of international media and NGOs from Port-au-Prince to the United States. The destruction in Kathmandu contributed to leave the Goudougoudou behind in people's minds.

By 2021, only a fraction of the reconstruction objectives was achieved. The reconstruction of Port-au-Prince is the case of an unfinished task, a process that generated much attention but failed to deliver long-lasting results. It is the case of an incomplete reconstruction in which stakeholders abandoned their beneficiaries and failed to honor their promises and reach the established goals (Lizarralde, 2021).

The problems started early on. Six months after the earthquake, the Haitian government reported to the United Nations that only 2% of the promised aid had been effectively delivered (ECOSOC, 2010). With time, problems accumulated: Objectives were not met, deadlines were postponed, projects were abandoned, promised funds were never delivered, and attention dwindled. A few ribbons were cut here and there: A housing project, called Lumane Casimir, was built by the Haitian government and some international agencies in the outskirts of the capital; new schools, and a few parks and hospitals were inaugurated in its metropolitan area. But Port-au-Prince went on with most of its problems unaddressed. By 2018 and 2019, Haitians had become exasperated, and a series of strikes and violent protests paralyzed the country (Lambert, 2019). Part of the frustration was related to the way the government managed the funds of PetroCaribe, a funding program resulting from a cooperation agreement with Venezuela. The situation became so critical that locals argued that the country had become "locked" (*peyi lòk*) and inoperative (Sénat, 2020).

What happened? Or rather, why were all the promises described in the Sendai Framework not effective in Haiti? There are perhaps two possible answers to this question: One is that the world overlooked Haiti and the objectives of the Sendai Framework were not adopted in the country. The other possible answer is that the principles of the Sendai Framework were applied but failed to achieve their objectives. I will argue here that both situations occurred. Several international stakeholders quickly forgot about Haiti, but also those who intervened in the country before and after 2015 made several mistakes that can be linked to the premises of the Sendai Framework. Some of those stakeholders were involved in the reconstruction of central districts in Port-au-Prince, others in a new area called Zoranjé, and others in a place with a Biblical name: Canaan.

3. Investing in disaster risk reduction in Port-au-Prince

Three ideas were popular in Port-au-Prince after the 2010 disaster and contributed to the failure of the Haitian reconstruction process. The first one was the idea of replacing whatever was left after the disaster with new "solutions." After the earthquake, several local and international experts reminded everyone that the Haitian capital sits on a fault line, which exposes it to tsunamis and earthquakes. Being in the passage of seasonal tropical storms and hurricanes, the city is also exposed to heavy rains, water surges, and floods. Partially built on the mountains and around water streams, the city is prone to violent water runoffs and landslides. Besides, with a large part of its development on the seashore, the city is increasingly exposed to climate-change-induced sea-level rise. Others claimed that Port-au-Prince was "the city of anarchy," where residents do as they please (Laënnec, 2015), and pollution, traffic, and lack of infrastructures made it impossible to reach decent living standards. Given all these arguments, journalists, engineers, and some scholars contemplated the idea of relocating the city of three million people to a safer location.

Of course, this ambition of replacing one dirty, wasted, and inefficient city with a brand-new one never happened. But innovators still saw in Haiti the opportunity to design from scratch sustainable and resilient prototypes of virtually everything. One of those early ideas was Harvest City, a new district designed to be built on the ocean by Schopfer Associates LLC, a US architecture and planning firm. The plan included hurricane-resistant floating units for 30,000 Haitians, and floating farms, harbors, markets, and schools (Schopfer Associates LLC, 2011).

Understandably, neither investors nor politicians took seriously Harvest City. The idea of building a Port-au-Prince neighborhood on the water was as absurd as unfeasible. Nonetheless, the idea of replacing the *cités*—Haitian slums and low-income neighborhoods—with new ones did take hold. And the objective of replacing existing constructions and institutions with new ones was quickly adopted by many international entities as a principle of urban intervention in Haiti. A common argument in 2010 was that a "refounding" of the Haitian State was required to save the country from total chaos (Bornstein et al., 2013; Geffrard, 2019). Existing institutions had to be dismantled and replaced with new, more democratic, and sustainable ones. According to Kasia Mika, a scholar who examined the narratives adopted after the Haitian disaster, several ideas at the time focused on the "renewal rather than recovery" of existing institutions (Mika, 2018, p. 44).

A second idea that also became popular after the disaster was that it was the absence of planning that prevented the country from thriving. International observers and NGO representatives often argued that the lack of "proper" urban and strategic planning constantly diminished the efficiency of infrastructures and services, and halted investment. A very common argument among international agencies, charities, and private donors was that many of the country's problems lied in insufficient urban planning, regulations, data, and studies. Foreign urban consultants and reports were needed to solve the city from total disruption. A proper cadastral plan was needed to unlock housing construction. Without reliable data about land property, it was impossible to develop urban initiatives and reach international standards. A 2017 World Bank report on Haiti argued that the country should move "from reconstruction to resilient urban planning for a bright future" (Lozano-Gracia & Garcia Lozano, 2017, p. 61). But the studies conducted by our research team at Université de Montréal and Université du Québec à Montréal from 2010 to 2017 demonstrated the existence of a different reality. We found that, over the past 20 years, there were hundreds of urban plans, regulations, and studies about Port-au-Prince. More than 200 studies centered on urban issues, over 10 policy documents were written by the government and agencies since 2010, and there were more than 37 urban plans for Port-au-Prince alone. And even when cadastral information was available, rules were almost never enforced. We eventually concluded that the problems of Port-au-Prince did not stem from a lack of planning or data, but insufficient implementation and enforcement capacity. There are sufficient laws and regulations in the country, but they are rarely enforced, and policies and plans are almost never implemented.

One of the projects that aimed at replacing central areas of the city with new constructions was led by Jean-Yves Jason, then mayor of Port-au-Prince. His plan for the city, designed by the IBI Group, a Canadian urban-design firm, proposed that a great part of the central district be demolished and replaced with modern buildings, mostly for government agencies and offices. The port would also be expanded and modernized, and green boulevards would connect the shore with the new government facilities. A similar initiative to refurnish the central district was led by the Prince Charles Foundation. Here the plan was to demolish several blocks of downtown Port-au-Prince (including some historic buildings) and replace them with New Urbanism-style buildings, plazas, and parks. Neither Jason's project nor the Prince Charles Foundation's initiative or many other projects designed after 2010 were implemented. They were all based on ideal plans, not effective capacities for navigating the challenges of securing funding and political support, and managing implementation.

Finally, the third idea was that new institutions were required in Haiti. In 2009, the Comité Interministériel d'Aménagement du Territoire (CIAT), an interministry committee, was created to deal with urban and strategic-planning issues. After the destruction, the Unité de Construction de Logements et de Bâtiments Publics (UCLBP), a new public body to deal with housing and reconstruction projects was also created. Even though the CIAT was in its infancy in 2010, there was an interest in creating additional institutions after the earthquake. One of them was led by the

International Association of French-speaking Mayors (AIMF), an organization of cities from about 52 countries, mostly former French colonies. The AIMF's idea was to create a new association of municipalities. Representatives of the AIMF ignored that such associations already existed, and the problem was that they were severely underfunded. Another similar idea was promoted by the European Union. Here the objective was to create a metropolitan institution, capable of overcoming the political and administrative fragmentation that characterizes the area of Port-au-Prince (where there are several municipal jurisdictions). Even though the CIAT and the UCLBP became key players after 2010, the new municipal institutions were never fully implemented.

4. Building back better in Haiti

In 2009, former US president Bill Clinton was appointed as the United Nations' special envoy for Haiti. Clinton had experience in postdisaster diplomacy. He had served as the UN Special Envoy for Recovery following the 2004 Indian Ocean earthquake and tsunami. After the Goudougoudou, Clinton and his wife joined forces with former president George W. Bush to create the Clinton Bush Haiti Fund. The Clinton Foundation led efforts to increase foreign investment in Haiti as well as a reconstruction program called Building Back Better. Clinton's efforts aimed at combining housing reconstruction with economic development. At the time, an advisor of the Clinton Global Initiative argued that "sustainable reconstruction can expand opportunity by investing in local entrepreneurs and supply chains" (Hendricks et al., 2010, p. 94). According to Clinton, the Building Back Better initiative would "lead to whole new industries being started in Haiti, creating thousands and thousands of new jobs and permanent housing" (The Gazette, 2011). Then he added: "This is the moment that holds the most promise for Haiti to overcome its past, once and for all" (Clinton, 2010, p. 4).

At the core of the plan was the objective of building a model district in Zoranjé, an empty area in the outskirts of Port-au-Prince (see Fig. 1). To build this model "satellite city," the Foundation mandated Malcom Reading Consultants, an English firm, to organize an international housing competition (Malcom Reading Consultants, 2010). Partners included the World Bank, the Inter-American Development Bank, Deutsche Bank, John McAslan + Partners (a British architecture firm), Arup (a multinational construction company), and the NGOs Architecture for Humanity and Habitat for Humanity International. A housing exhibition to

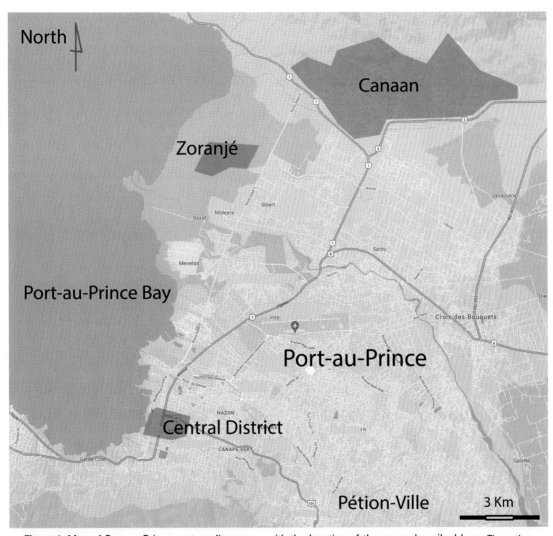

Figure 1 Map of Port-au-Prince metropolitan area, with the location of the areas described here. The author.

display the best housing prototypes submitted was then built in
Zoranjé. It is estimated that the exhibition cost more than USD2
million—including USD500,000 from the Clinton Foundation
(Sullivan & Helderman, 2015).

The idea was that private developers and citizens would visit
the prototypes and select the solutions they wanted to buy or
build. The houses were all single-story detached units, to be
sold for a price ranging from USD21,000 to USD70,000. The prototypes included units made of metallic containers, prefab and

plastic panels, and recycled materials. The urban study (funded by the Clinton Foundation, Deutsche Bank, and Digicel, a Haitian telecommunications firm) was conducted by the Graduate School of Design at Harvard University and MIT's School of Architecture and Planning. The study recommended "a strategy of private sector involvement and economic development," as well as "strategic partnerships with Haitian banks" (Harvard University Graduate School of Design Social Agency Lab & MIT Department of Architecture and Planning, 2011, p. 155). The researchers from Harvard University and MIT found that Zoranjé was a flood-prone area. But this did not deter the MIT and Harvard professors from going ahead with the urban plan for the model district.

The houses were too expensive, and Zoranjé lacked basic services and transportation. Understandably, the place never attracted local businesses or residents. The master plan designed by specialists at MIT and Harvard was never developed. Apart from the original prototypes, none of the houses of the Building Back Better program were built. The jobs and economic activities promised by Clinton's initiative did not materialize either. The Clinton Foundation and the American urban planners and designers left Haiti a few years after the exhibition's construction. The housing exhibition in Zoranjé became a ghost town... in a flood-prone area.

5.   Investing in disaster risk reduction in the promised land

Following the earthquake, a group of nearly 60 affected families squatted in a golf course in Pétion-Ville, a wealthy neighborhood in the capital city. NGOs developed a project to relocate them to Corail, an unoccupied area in the northern mountains of the city (Mozingo, 2016). Corail lacked roads, infrastructures, and public services, but this did not stop charities from building temporary shelters (mostly made of wood and corrugated sheets). The place quickly attracted more and more disaster survivors; as it grew, it came to be known as Canaan—the Biblical promised land.

A few months later, the president of Haiti, René Préval, issued a bill that declared Canaan an area for public use. With this bill, Préval effectively expropriated the owners of the land and legitimized its occupation by disaster survivors. More and more people got attracted by the dream of becoming landowners in Canaan. Today, it is estimated that the place is occupied by more than 300,000 people. Almost everything in Canaan (including homes, infrastructures, and services) has been built by local residents without the participation of government, urban planners, architects, engineers, or formal construction companies. But Canaan is not just another Port-au-Prince slum. Compared with notorious informal settlements such as Cité Soleil or Cité Lajoie, Canaan is a functional "city" (it is also one of Haiti's largest urban centers).

Technically, the place is not (yet) a city. But residents, land mafias, and informal construction companies built not only houses but also roads, parks, churches, schools, and water wells. Today, it is home to both low-income and middle-class families. And an informal real-estate market is thriving in Canaan. According to Anne-Marie Petter, a Canadian scholar who has studied the settlement since 2010, land prices in some areas of Canaan have risen as much as 20 times over.

By 2015 the place had, understandably, attracted attention from charities and international agencies. Authorities at the UCLBP, the CIAT, and charities worried that, as it grows, the settlement deteriorates and follows the path of Port-au-Prince's *cités*. One of those NGOs was the American Red Cross, which had provided health services in the area since the emergency phase. Between 2015 and 2018, it invested USD21 million to upgrade Canaan's infrastructures and public space. With the help of the UCLBP, this American NGO set up the goal of developing Canaan "in accordance with urban planning standards and principles, and in keeping with the international vision of urban development" (UCLBP, 2015).

The process included setting up a governance structure based on local leaders and neighborhood committees (known as *tables de quartier*). For urban planning, the American Red Cross partnered with UN-Habitat. For a couple of years, UN-Habitat worked in designing a master plan for Canaan that complied with sustainable-development principles. To create this plan, UN-Habitat, the American Red Cross, and the UCLBP organized a series of participatory workshops with local leaders and representatives of public and private organizations. The process led to a series of recommendations and drawings for "proper" urban design in Canaan. The plan included large paved roads with sidewalks and lighting.

The UN-Habitat plan seems appropriate at first glance. But anyone who knows how Port-au-Prince's public space works can figure out that it is completely decontextualized. The streets and roads of Port-au-Prince and most Haitian cities are not only for cars and buses. They are the place for street vendors, pedestrians, animals, social gatherings, sport activities, and children's games. But the Canaan master plan was designed according to traditional Western urban standards. The streets followed international urban-planning regulations but seemed disconnected from the Haitian urban reality. More worryingly, the plan would significantly increase pavement, and thus watertight surfaces, which would facilitate water runoff during rains and reduce the possibility that water percolates through the ground and reaches underground wells.

At this point, you might be worried that Canaan's master plan will bring additional problems to the area. But don't worry too much. Petter and colleagues consider that "expanding the road network alone, from 600 km to 1500 km," as detailed in the UN-Habitat plan, "would cost an estimated US$1.5 billion, or 81% of Haiti's national budget for 2018–2019" (Petter et al., 2020). Given these numbers, it is not surprising that very little of this master plan has been executed. The American Red Cross completed the construction of only a few parks and a small fraction of roads have been paved. Like many other plans to invest in disaster risk reduction for resilience, the Canaan master plan has been shelved.

6. Lessons learned from the Haitian case

The Haitian case exemplifies several contemporary problems with disaster risk reduction and reconstruction—many of them are closely linked to the premises of the Sendai Framework. At first glance, it is difficult to see how the Sendai Framework can contribute to repair all the injustices and pain that happened in Haiti in the past decade. Finally, the framework is a recital of good intentions. What can be wrong with that?

Well, like many other aspects of human life, the problem lies not on what we *say*, but on what we *do* with words. Vague terms and objectives written in institutional jargon can be interpreted in many ways—and it is often the interpretation that fits the objectives of the most powerful that is adopted. It is rarely the interpretation that fits the needs and expectations of the vulnerable, the excluded, the marginalized, and the poor that is put forward.

It might be tempting to argue that the Sendai Framework is like a knife that can be used for good and bad purposes. Following this line of thought, it is the way the framework was adopted in Haiti that was the problem, not its premises themselves. This might be the case. But I contend here that the framework's vagueness and empty jargon facilitates the type of "wrong interpretations" that led to several problems in Port-au-Prince.

Finally, there is no single mention in the framework of the type of root causes that led to the Goudougoudou: neoliberalism, colonialism, imperialism, elitism, savage capitalism, corruption, cronyism, partisanship, and racism. Unfortunately, Haiti is not a rare exception. This book shows that these are the root causes of many other disasters in the Philippines, Nigeria, Sri Lanka, China, Turkey, Nepal, and Indonesia.

So how does a framework end up condoning these social injustices that lead to disasters? The most obvious answer is the decision to mention neither these injustices nor practical examples that can resonate with local leaders and decision-makers (this, of course, would be too controversial for a UN document). But there are also more subtle ways in which frameworks act in complicity with the manipulation and abuses of apparently virtuous goals. Here I argue that there are six ways in which vaguely written objectives are affecting DRR and reconstruction today.

First, an overreliance on international consultants. The Sendai Framework argues that "the United Nations and other international and regional organizations, international and regional financial institutions and donor agencies engaged in disaster risk reduction are requested, as appropriate, to enhance the coordination of their strategies" (UNISDR, 2015, p. 25). It is therefore not surprising that the DRR and climate-change-adaptation agendas today have been presented as a resilience crusade by the Rockefeller Foundation, ICLEI, UN-Habitat, UNDP, C40, and other mammoths of the urban-consulting world. This approach is also too optimistic about the advantages of international aid and the role of international financial institutions. The document calls for "international financial institutions, such as the World Bank and regional development banks, to consider the priorities of the present Framework for providing financial support and loans for integrated disaster risk reduction to developing countries" (UNISDR, 2015, p. 26). According to the Sendai Framework, DRR requires "an enhanced provision of means of implementation, including adequate, sustainable and timely resources, through international cooperation and global partnerships for development, and continued international support" (UNISDR, 2015, p. 24).

These goals have failed to address the secondary effects and downsides of aid (Lizarralde, 2019). The consequence of this overreliance on international consultants and money in Haiti and many other places is that vast amounts of resources are invested in writing reports, diagnostics, frameworks, guidelines, pathways, guides, standards, and other documents that are based on foreign knowledge and perceptions, but fail to understand the specificities of local contexts. The preparation of these documents allows international agencies to show goodwill and gain visibility, but consumes resources that could be devoted to invest in local professionals, researchers, and practitioners who know best the realities of the place.

International consultants of UN agencies and private charities, such as the Rockefeller Foundation and the Clinton Foundation, typically retort that the plans and initiatives they support are not necessarily "theirs." The argument is that they "simply" partner with local institutions to design the plans and write the policy documents (NGOs in Haiti, for instance, are always careful of adding the Haitian government's logo in all reports). But international organizations were more than advisers in Haiti. In many cases, they were the de facto leaders behind strategic decisions. They often attempted to shortcut the corruption and cronyism that characterize municipalities and the central government in Haiti.

But the Haitian case shows that relying on international institutions to design urban plans, policies, or new institutions is hardly the answer to local problems. Without implementation capacity, these efforts lead to reports, plans, and documents that are shelved and abandoned. Furthermore, the strategy of hiring international consultants hinders the possibility that local architects, engineers, and urban planners participate in change. In Haiti, local professionals were often invited as collaborators and advisors. But without decision-making power and respectful contracts for their expertise, they often became spectators of the changes proposed by international agencies and charities. This was a significant problem in Haiti. According to the Centre for Economic and Policy Research, USAID (the American agency for international development) awarded USD2.3 billion to reconstruction activities in Haiti. Yet, more than 55% of resources were awarded to organizations and firms located in the United States. It is estimated that only 2.3% of the resources went directly to Haitian companies or organizations (Johnston, 2018).

Second, an exaggerated optimism on the introduction of new jargon. According to a World Bank report on Haiti, the country has a "long history of resilience in the face of slavery, colonialism, political oppression, widespread destruction from natural hazards, social exclusion, inequality, and poverty" (Lozano-Gracia & Garcia Lozano, 2017, p. 2). This past, the report argues, "determines its current challenges to development, but most importantly, the opportunities that lie ahead" (Lozano-Gracia & Garcia Lozano, 2017, p. 2). This World Bank report is not an exception. In fact, recent frameworks, including Sendai, have helped universalize a narrative of risk and disasters based on the resilience and sustainable-development jargon.

The main problem is that these narratives not always reflect the real needs and expectations of people affected by disasters. Recent studies have shown that ideas of social change, social injustice, social status, and aspirations are more relevant than those of resilience jargon for low-income and historically marginalized populations in Haiti (Petter et al., 2020), Colombia (Lizarralde et al., 2020; Muñoz et al., 2019; Páez et al., 2019), Cuba (Aragon-Duran et al., 2020; Lizarralde et al., 2015), and other places. After many years working in Haiti, members of my team and I had rarely heard local residents talk about "resilience," "adaptive capacities," or "climate adaptation." Quite often, however, they have told us about political corruption, exclusion, elitism, oppression, and other social and environmental injustices. Similarly, local authors have also emphasized the problems of political dishonesty (Darius, 2019), patriarchal structures (Organisation des Femmes Haïtiennes, 2019), exclusion (Théodat, 2013), and exploitative forms of urban development (Alphonse, 2015).

The resilience and sustainability jargon has become an international set of prefabricated ethical reasoning that is often disconnected from the realities of the places. In this way, it helps propose solutions that are ill-adapted in many places—and particularly in the Global South.

Third, an emphasis on public–private partnerships. Ever since the publication of the Brundtland report on sustainable development (Brundtland, 1987), United Nations agencies have insisted on the importance of creating alliances between governments and companies to deal with environmental degradation, risks, and disasters. The idea of partnered action seems not wrong at first glance. But empirical research demonstrates that it has helped legitimize dubious relationships between political and economic elites (Lizarralde, 2021). Partnerships between politicians and mighty companies are one of the main causes of environmental degradation in the Amazon Region. These alliances facilitate the construction of periurban developments that put rural peasants and traditional villages at risk in Vietnam and other countries, and they perpetuate corruption and cronyism in Haiti and many other places.

The Building Back Better initiative in Haiti was developed around the idea of planning solutions through close alliances between donors, government, and private firms such as local and international banks. But these alliances rarely focused on the needs of the most vulnerable, failed to tackle political opposition and bureaucracy, and seldom materialized in implementation plans. More worryingly, they aimed not at reinforcing the capacity of the Haitian State but to bypass it by delegating work to the private sector. When NGOs and firms withdrew, there was little left in the government structure to deal with social needs and implementation objectives.

Fourth, an emphasis on objectives, but not on means and implementation. Today, thousands of cities and territories have adopted risk-reduction plans or have integrated DRR in existing strategic and planning tools. We must celebrate the inclusion of risks and disasters in policy and planning. But in most cases, efforts have been placed on determining noble objectives, without an appropriate allocation of resources, competent staff, and adminis-trative structures and funding. The consequence in Haiti and many other places is that there are numerous plans, strategic documents, white papers, laws, and regulations dealing with risk, environ-mental degradation, and disaster prevention—many of which are written by consultants of the Rockefeller Foundation, ICLEI, UN-Habitat, UNDP, C40, and other similar institutions. But there are very few implementation programs and very little real action on the ground. One of the reasons is that once international institu-tions have published their reports and gained visibility in a certain location, they move on to other places to start new consulting man-dates. In this way, these institutions devote little efforts in guaran-teeing continuity, implementing measures, consolidating change, navigating the difficult decisions required in action, and dealing with the political debates that are required in policy. Another reason is that international frameworks, often written in terms of desirable goals and not political changes, perpetuate the lack of emphasis on the type of implementation that is badly needed. The third reason is that Sendai and other frameworks fail to address the conflicts and tensions that often emerge when several virtuous objectives must be achieved at the same time. These frameworks rarely tackle the secondary effects, blind spots, and un-intended consequences of the solutions found to a given problem.

A common problem is the interest in creating new institutions. International consultants know how to create new committees, agencies, and bodies. But they are rarely focused on how these new institutions will survive. Quite often, they become empty shells with little capacity to enforce regulations or deploy implantation.

Fifth, an emphasis on technology. The plans for Zoranjé were based on the idea that the housing problem in Haiti was one of lack of appropriate construction technology. Builders and developers rushed to develop prefab and industrialized solutions to be able to build housing quickly. But most failed to realize two problems. First, that foreign construction technologies and solutions were a poor fit for the local climate and conditions, and the needs and expectations of residents. Second, that these technologies did not tackle the real bottlenecks of housing development in Haiti, including the difficulty to legalize and obtain serviced land, problems to obtain permits, lack of skilled labor, and insufficient infrastructures to connect homes to basic services.

"It is crucial to enhance technology transfer," argues the Sendai Framework (UNISDR, 2015, p. 24). The framework embarks on a current trend (ever more common as we realize the effects of climate change) in both developed and developing nations: a focus on technology to solve environmental challenges and reduce risks. Recent studies have shown that climate change adaptation, for instance, has largely focused on the "hazard" component, with little efforts to address racism, colonialism, imperialism, neoliberalism, segregation, neglect, and other social injustices that lead to disasters (Lizarralde, 2020). While many argue that it is patterns of exclusion and segregation that need to be changed to reduce disasters (Lizarralde, 2021), international frameworks have consistently contributed to a blind optimism on technology.

Sixth, an emphasis on a superficial approach to "community" participation and empowerment. According to the Sendai Framework, DRR requires "empowerment and inclusive, accessible and nondiscriminatory participation, paying special attention to people disproportionately affected by disasters, especially the poorest" (UNISDR, 2015, p. 13). This is a noble objective that is difficult to disagree with. But the Haitian case and other empirical studies (Barenstein & Leemann, 2012; Lizarralde, 2021) show that even UN agencies often resort to superficial interpretations of what citizen participation should be. The public is frequently manipulated to legitimize decisions taken in advance, people are consulted about predetermined ideas, public audiences are routinely used to gain support to ideas and projects that do not benefit the poor and marginalized, and little voice and decision-making power are typically given to citizens and segregated social groups. "Participation" has become the most common term to provide a thin coat of ethical responsibility to all DRR and climate-adaptation initiatives.

7. Where and how to invest for disaster risk reduction?

International frameworks in general, and those sanctioned by UN agencies in particular, are written to create the sense that there is consensus (or rather, a form of unanimity), not to explore reality. But disaster risk reduction, environmental protection, and reconstruction are political actions. As such, they create winners and losers and determine the distribution of resources among competing interests and expectations. Decisions taken in the face of risk and environmental degradation affect cities and territories forever and almost always cause secondary effects, unintended consequences, and long-term impacts. But frameworks (by their very definition), instead, are written to create consensus, flatten controversies, and avoid debates. They lead us to believe that we agree on the objectives ahead, that goals are always virtuous, and that there is no need to worry about contradictions and conflicting interests.

The problem is that objectives are interpreted by political and economic elites in ways that fit their own agendas. Thus, they often fail to address the needs and expectations of those that the frameworks want to help. In other words, and paraphrasing Ben Wisner, they fail to deal with the root causes of disasters.

Frameworks and guidelines written in vague language often fail to engage in the type of political debate that is needed today. They provide visibility to powerful international institutions staffed with consultants. But they do very little to facilitate the type of political discussion that is required to change behaviors, policy, and practices on the ground.

Hyogo, Sendai, and other UN-supported frameworks have become empty shells. They are full of the right jargon but not politically engaged. As such, they are of little use today to understand risks and respond to disasters. But people (even those in poor countries) use them for fear of "missing the boat." Should we (scholars) perpetuate or condone this empty—but sometimes heart lifting—jargon? I do not think so. I think academics must keep challenging these frameworks. We must show their limitations, blind spots, and oversights. Climate change and disasters are political processes. They must always be treated as such. Only in this way will we know where to invest in disaster risk reduction for change.

References

Alphonse, R. (2015). Construction: SOS des firmes haïtiennes [Construction industry: The SOS call of Haitian firms]. *Le Nouvelliste*. Retrieved from http://lenouvelliste.com/article/140377/Construction-SOS-des-firmes-haitiennes (in French).

Aragon-Duran, E., Lizarralde, G., González-Camacho, G., Olivera-Ranero, A., Bornstein, L., Herazo, B., & Labbé, D. (2020). The language of risk and the risk of language: Mismatches in risk response in Cuban coastal villages.

International Journal of Disaster Risk Reduction, 50, 101712. https://doi.org/
10.1016/j.ijdrr.2020.101712

Barenstein, J. D., & Leemann, E. (2012). *Post-disaster reconstruction and change:
Communities' perspectives.* London: CRC Press.

Bornstein, L., Lizarralde, G., Gould, K. A., & Davidson, C. (2013). Framing
responses to post-earthquake Haiti: How representations of disasters,
reconstruction and human settlements shape resilience. *International
Journal of Disaster Resilience in the Built Environment, 4*(1), 43–57. https://
doi.org/10.1108/17595901311298991

Brundtland, G. H. (1987). *Our common future.* Report of the World Commission
on Environment and Development. New York, NY: United Nations.

Clinton, B. (2010). Our commitment to Haiti. *Innovations: Technology,
Governance, Globalization, 5*(4), 3–5. https://doi.org/10.1162/INOV_a_00037

Darius, D. (2019). L'opposition s'unit et se donne une alternative pour la
refondation de l'État à Mirebalais [Opposition parties unite in Mirebalais for
the refounding of the State]. *Le Nouvelliste.* Retrieved from https://
lenouvelliste.com/article/206683/lopposition-sunit-et-se-donne-une-
alternative-pour-la-refondation-de-letat-a-mirebalais (in French).

ECOSOC (United Nations Economic and Social Council). (2010). *Less than 2 per
cent of promised reconstruction aid for quake-devastated Haiti delivered,
Haitian Government envoy tells Economic and Social Council.* Retrieved from
https://www.un.org/press/en/2010/ecosoc6441.doc.htm.

Geffrard, R. (2019). Rasire Aity, pour la refondation d'Haïti [*Rasire Aity,* toward
the refounding of Haiti]. *Le Nouvelliste.* Retrieved from https://lenouvelliste.
com/article/197829/rasire-ayiti-pour-la-refondation-dhaiti (in French).

Hannah, R., & Roser, M. (2019). *Natural disasters.* Retrieved from https://
ourworldindata.org/natural-disasters.

Harvard University Graduate School of Design Social Agency Lab, & MIT
Department of Architecture and Planning. (2011). *Designing process:
Exemplar community pilot project Zoranjé, Port-au-Prince.* Boston: Kendall
Press.

Hendricks, B., Christensen, A., & Toussaint, R. (2010). Green reconstruction:
Laying a firm foundation for Haiti's recovery. *Innovations: Technology,
Governance, Globalization, 5*(4), 129–141. https://doi.org/10.1162/
INOV_a_00049

Johnston, J. (2018). *Where does the money go? Eight years of USAID funding in
Haiti.* Report of the Center for Economic and Policy Research, Haiti Relief
and Reconstruction Watch. Retrieved from https://cepr.net/where-does-the-
money-go-eight-years-of-usaid-funding-in-haiti.

Laënnec, H. (2015). Haïti: « L'habitat, c'est le chaos et l'anarchie » [Haiti: "Chaos
and anarchy in the housing sector"]. *Le Nouvelliste.* Retrieved from https://
lenouvelliste.com/m/public/index.php/article/140559/haiti-lhabitat-cest-le-
chaos-et-lanarchie (in French).

Lambert, R. (2019). Peyi lòk, « 3,7 millions de personnes sont en situation
d'insécurité alimentaire aiguë » [*Peyi lòk,* "3.7 million Haitians facing acute
food insecurity"]. *Le Nouvelliste.* Retrieved from https://lenouvelliste.com/
article/208791/peyi-lok-37-millions-de-personnes-sont-en-situation-
dinsecurite-alimentaire-aigue (in French).

Lizarralde, G. (2019). Does aid (actually) aid in avoiding disasters and rebuilding
after them? *OD Debates.* Retrieved from https://oddebates.com/8th-debate.

Lizarralde, G. (2020). Is adapting to climate change (really) our best choice? *OD
Debates.* Retrieved from https://oddebates.com/10th-debate.

Lizarralde, G. (2021). *Unnatural disasters: Why most responses to risk and climate change fail but some succeed.* New York, NY: Columbia University Press.

Lizarralde, G., Paéz, H., Lopez, A., Lopez, O., Bornstein, L., Gould, K., et al. (2020). We said, they said: The politics of conceptual frameworks in disasters and climate change in Colombia and Latin America. *Disaster Prevention and Management, 29*(6), 909–928. https://doi.org/10.1108/DPM-01-2020-0011

Lizarralde, G., Valladares, A., Olivera, A., Bornstein, L., Gould, K., & Barenstein, J. D. (2015). A systems approach to resilience in the built environment: The case of Cuba. *Disasters, 39*(s1), s76–s95. https://doi.org/10.1111/disa.12109

Lozano-Gracia, N., & Garcia Lozano, M. (Orgs.). (2017). *Haitian cities: Actions for today with an eye on tomorrow.* Washington, DC: World Bank Group. Retrieved from http://documents.worldbank.org/curated/en/709121516634280180/pdf/122880-V1-WP-P156561-OUO-9-FINAL-ENGLISH.pdf.

Malcom Reading Consultants. (2010). *Building Back Better communities Port-au-Prince, Haiti: Request for proposal on behalf of the Government of the Republic of Haiti.* London: Malcom Reading Consultants.

Mika, K. (2018). *Disasters, vulnerability, and narratives: Writing Haiti's futures.* New York, NY: Routledge.

Mozingo, J. (2016). Sean Penn's hands-on aid for Haiti quake victims an earlier sign of his risk-taking. *Los Angeles Times.* Retrieved from https://www.latimes.com/local/lanow/la-fg-sean-penn-haiti-20160110-story.html.

Muñoz, L., Paéz, H., Lizarralde, G., Labbé, D., & Herazo, B. (2019). Adaptation to water scarcity: Water management strategies led by women on the Caribbean island of San Andres. *Trialog, 134*(3), 14–18.

Organisation des Femmes Haïtiennes. (2019). Lettre ouverte d'organisations de femmes aux dirigeants des trois pouvoirs de l'État haïtien et aux leaders de l'opposition [Open letter from women's organizations to the leaders of the Haitian State and the opposition]. *Le Nouvelliste.* Retrieved from https://lenouvelliste.com/article/198441/lettre-ouverte-dorganisations-de-femmes-aux-dirigeants-des-trois-pouvoirs-de-letat-haitien-et-aux-leaders-de-lopposition (in French).

Páez, H., Díaz, J., Lizarralde, G., Labbé, L., & Herazo, B. (2019). Coping with disasters in small municipalities: Women's role in the reconstruction of Salgar, Colombia. *Trialog, 134*(3), 9–13.

Petter, A.-M., Labbé, D., Lizarralde, G., & Goulet, J. (2020). City profile: Canaan, Haiti—A new post disaster city. *Cities, 104*, 102805. https://doi.org/10.1016/j.cities.2020.102805, 2020.

Schopfer Associates LLC. (2011). *Harvest City.* Retrieved from https://www.youtube.com/watch?v=srwFzw82o6w.

Sénat, J. D. (2020). Quand Jovenel Moïse compare instabilité politique, « peyi lòk » et le séisme de 2010 [The impacts of political instability, "*peyi lòk*", and the 2010 earthquake, as compared by Jovenel Moïse]. *Le Nouvelliste.* Retrieved from https://lenouvelliste.com/article/211075/quand-jovenel-moise-compare-instabilite-politique-peyi-lok-et-le-seisme-de-2010 (in French).

Smith, A. (2020). *2010-2019: A landmark decade of U.S. billion-dollar weather and climate disasters.* Retrieved from https://www.climate.gov/news-features/blogs/beyond-data/2010-2019-landmark-decade-us-billion-dollar-weather-and-climate.

Sullivan, K., & Helderman, R. (2015). How the Clintons' Haiti development plans succeed—and disappoint. *The Washington Post.* Retrieved from https://www.washingtonpost.com/politics/how-the-clintons-haiti-development-plans-succeed–and-disappoint/2015/03/20/0ebae25e-cbe9-11e4-a2a7-9517a3a70506_story.html.

The Gazette. (2011). *Haiti: After the quake*. Retrieved from http://www.montrealgazette.com/news/haiti-quake/index.html.

Théodat, J.-M. (2013). Port-au-Prince en sept lieues [Port-au-Prince: Relocate or regenerate?]. *Outre-Terre, 1–2*(35–36), 123–150 (in French).

Thomas, F. (2020). Comment Haïti est devenu la « république des ONG » [How Haiti became the "republic of NGOs"]. *Le Monde*. Retrieved from https://www.lemonde.fr/idees/article/2020/01/09/comment-haiti-est-devenu-la-republique-des-ong_6025258_3232.html (in French).

UCLBP (Unité de Construction de Logements et de Bâtiments Publics). (2015). *Rencontre de présentation du plan d'action pour la restructuration urbaine de la zone de Canaan et ses environs [Meeting for the presentation of the urban-restructuring action plan of Canaan and its surroundings]*. Retrieved from https://www.facebook.com/notes/uclbp/rencontre-de-pr%C3%A9sentation-du-plan-daction-pour-la-restructuration-urbaine-de-la-/732019023591228 (in French).

UNISDR (United Nations Office for Disaster Risk Reduction). (2015). *Sendai Framework for Disaster Risk Reduction 2015–2030*. Retrieved from https://www.undrr.org/publication/sendai-framework-disaster-risk-reduction-2015-2030.

Wisner, B. (2020). Five years beyond Sendai—Can we get beyond frameworks? *International Journal of Disaster Risk Science, 11*, 239–249. https://doi.org/10.1007/s13753-020-00263-0

A

Investing in design for disaster risk reduction

Investing in community participation for disaster recovery

Arlene Christy D. Lusterio[1], Geraldine R. Matabang[1] and Ma. Theresa Amor J. Tan Singco[2]

[1]*TAO-Pilinas (Technical Assistance Organization), Inc., Quezon City, Metro Manila, Philippines;* [2]*Disaster Risk Reduction Network Philippines (DRRNetPhils), Quezon City, Metro Manila, Philippines*

1. Sitting astride the ring of fire and typhoon path

The Philippines is the world's fourth most susceptible country to climate change, according to the Long-Term Climate Risk Index spanning 1999–2018, and ranks second in the 2018 Global Climate Risk Index (Eckstein, Künzel, Schäfer, & Winges, 2019). True to these risk indicators, the Philippines has undergone many major disasters due to various hazards in the past two decades. Each disaster is a new experience and a tragic reminder of how much we lack preparedness and how poorly we manage disaster risks. These disasters should have given us lessons on how to deal with them proactively.

The Philippines is a tropical country located in Southeast Asia. This 300,000-km^2 archipelago has 7641 islands and a 36,289-km long coastline (Government of the Philippines, 2021). Approximately 2000 islands are inhabited, constituting three groups: Luzon in the north and west, Visayas in the center, and Mindanao in the south (Boquet, 2017). The eastern seaboard, the most susceptible to typhoons and climate change, is classified as type II climate zone (Fig. 1.1). The Philippines' 2011 climate-change exposure map (Fig. 1.2) shows that the whole country is exposed to sea-level rise, which is being experienced at a rate thrice the global average (Government of the Philippines, 2020). It also

Investing in Disaster Risk Reduction for Resilience. https://doi.org/10.1016/B978-0-12-818639-8.00016-8

25

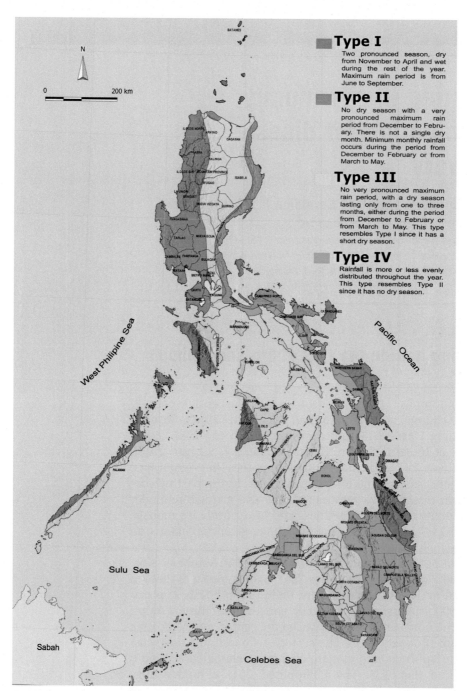

Type I

Two pronounced season, dry from November to April and wet during the rest of the year. Maximum rain period is from June to September.

Type II

No dry season with a very pronounced maximum rain period from December to February. There is not a single dry month. Minimum monthly rainfall occurs during the period from December to February or from March to May.

Type III

No very pronounced maximum rain period, with a dry season lasting only from one to three months, either during the period from December to February or from March to May. This type resembles Type I since it has a short dry season.

Type IV

Rainfall is more or less evenly distributed throughout the year. This type resembles Type II since it has no dry season.

Figure 1.1 The Philippines' four climate zones, based on the Modified Coronas Classification. From PAGASA (Philippine Atmospheric, Geophysical and Astronomical Services Administration). (2014). *Climate map of the Philippines.* Retrieved from http://bagong.pagasa.dost.gov.ph/information/climate-philippines.

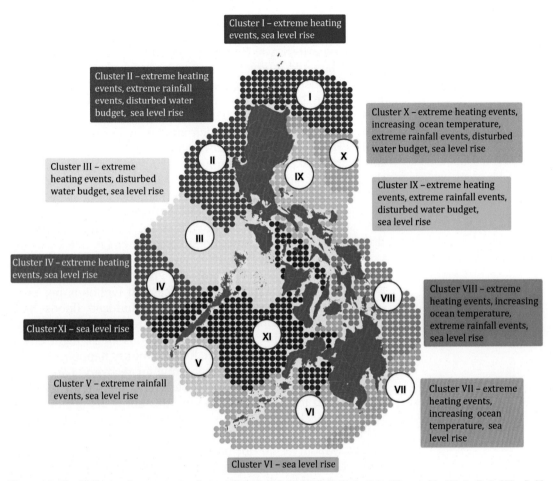

Figure 1.2 The Philippines' exposure to climate change. From David, L. T., Borja, R. T., Villanoy, C.L., Hilario, F., & Aliño, P. M. Developing a Philippine climate-ocean topology as input to national vulnerability assessments. In A. M. Lagmay (Ed.), *Proceedings of the Asian Association on Remote Sensing* (p. 3), New York, NY, Curan Associates Inc., 2012.

shows that the eastern seaboard is prone to increasing ocean temperatures, extreme heat and extreme rainfall events, and disturbed water budget. Such exposure further increases the already high precipitation level in the region.

The Philippines had a population of around 101 million people in 2015, growing at an average rate of 1.7% per year from 2010 to 2015. The population is highly concentrated in Luzon (57%) (PSA, 2017). About 60% of the Philippine cities and 56% of municipalities are located along the coasts, where about 52% of the urban population (and 62% of the total population) live. Considering that climate-change impacts are strongly felt in the coastal areas, most Filipinos are thus exposed to climatic hazards. Moreover, most poor and vulnerable people live as informal settlers along waterways. Hence, the possibility of a disaster is high without targeted interventions.

Furthermore, the Philippines is situated in the Pacific Ring of Fire, having 24 active volcanoes and 5 active fault traces that generate an average of 20 earthquakes per day (or 7300 per year), about 100 to 150 of which are felt by humans (ABS-CBN, 2019; PHIVOLCS, 2021). In 2019 alone, the country experienced four earthquakes with magnitudes ranging from 5.4 to 6.9 on the Richter scale (PHIVOLCS, 2020). Volcanic eruptions also left a very strong imprint in the Filipinos' memories. The cataclysmic 1991 Mount Pinatubo eruption in Central Luzon erased a significant portion of the province of Pampanga. Until now, lahars that flowed out of Mount Pinatubo still clog drainage lines, causing flooding in many places in the region.

But this is not the full picture. The Philippines also sits astride the typhoon path, experiencing at least 20 typhoons per year, about 5 of which are destructive. Rain-induced floods have affected entire regions, inundated dense resettlement sites, and caused landslides in rural and mountainous areas. Eight of the ten costliest typhoons that have hit the country happened within the 2010s decade. There is hardly time for recovery. Large-scale displacement and damage to property and crops are caused by multiple hazards. This is becoming the norm. Its geologic and meteorological characteristics make the Philippines very exposed to hazards. Filipinos have little choice but to live with these conditions, which yet do not necessarily need to engender massive disasters. However, the prevailing socioeconomic conditions of the Filipinos living below the poverty line (16.6% of the population) make them vulnerable to hazards' impacts (PSA, 2019). Farmers, fisherfolk, and people settled in rural areas comprise 82% of this vulnerable population, and most of them also live in the most exposed sites (PSA, 2020).

This chapter looks into postdisaster responses and disaster risk reduction (DRR) efforts of some civil-society organizations (CSOs) in the Philippines and their contributions to building resilience. Specifically, in Section 2, we look into relevant endeavors by the National Government in support of the third priority of the Sendai Framework and how this trickles down to the community level. We finalize Section 2 with a description of our study's method. In Section 3, we present two case studies of postdisaster response, dissecting the applications of participatory approaches in planning, design, and construction, as well as its relevance in building resilience. We also discuss the factors contributing to effective participation. In Section 4, we present our analysis looking at the practices, structures, and linkages underlying these recovery processes. Regarding practices, we examine how knowledge transfer is made to ensure effective learning and how this is applied in the participatory processes. In structures, we scrutinize shelter models, design, and construction interventions, as well as the process of involving stakeholders. In linkages, we look into connections between the Sendai Framework, the Philippine laws as localized DRR translations, the stand of networks toward disaster risk reduction and management (DRRM), and the consortium approach for massive disaster response. In this same section, we further discuss the value of participation and localization, and how related endeavors contribute to building resilience. Finally, in Section 5, we summarize some investments in various forms of participation and propose these to be directed to localized actions and structural measures toward resilience.

2. Institutionalizing disaster risk reduction and management

The Philippine landmark legislations on DRR and on climate change have been globally recognized, serving as models to examine critical provisions for good DRR and adaptation policy linkages (Llosa & Zodrow, 2011). This, however, has not always been the case.

Improvements in the DRR law had been met by various conflicting bills, priorities, and proposed piecemeal changes on highly specific issues, and had been more focused on response (Benson, 2009). The country's policies, legislations, and institutional arrangements had remained reactive almost halfway into the Hyogo Framework for Action, whereas the attempts for legislative changes dragged on for almost a decade. In comparison, due to its urgency being recognized by legislators and political leaders, the Climate Change Act took only three years of legislative processes and was enacted in 2009. In the following year, the Philippine Disaster Risk Reduction and Management Act (RA 10121) finally replaced the 1978 Presidential Decree (PD) 1566, itself preceded by the Executive Order (EO) 335, issued in 1941 (Table 1.1). This earlier legislation (EO 335) established the Civilian Emergency Administration and multilevel emergency committees on civilian participation in preparation for World War II (COA, 2014).

Table 1.1 Key milestones toward the formulation of the RA 10121.

Year	Milestones	Provisions
1978	Presidential Decree 1566	— Established and mandated the National Disaster Coordinating Council (NDCC) as the highest policy-making body and focal organization for disaster management in the Philippines. The NDCC is composed of 18 government agencies and the Philippine Red Cross. — The NDCC has regional- and barangay-level counterparts (barangay is the Philippines' basic political subdivision and primary planning and implementing unit of government programs and basic services). The function of the regional and local councils includes ensuring risk-sensitive regional development plans and the integration of DRR and climate change adaptation (CCA) into local development plans, programs, and budgets.
2005	Adoption of the Hyogo Framework for Action by 168 states, including the Philippines	
	NDCC Four-Point Plan of Action on Disaster Preparedness	First national plan for disaster preparedness and mitigation focused on (1) improvement of early-warning systems, (2) public awareness, (3) involvement of the private sector, and (4) capacity building for local government units.
2007	Disaster Risk Management Master Plan of Metro Manila	— Recognizing disaster risk management (DRM) as a shared responsibility, the DRM master plan of Metro Manila (DRMMP) facilitated a cross-sectoral, interagency course of action. — The DRMMP engaged local stakeholders toward the elaboration of a DRM agenda consisting of legal, institutional, financial, social, and technical elements (EMI, 2007).
2009	Strengthening Disaster Risk Reduction in the Philippines: Strategic National Action Plan (SNAP 2009–2019)	Provided the framework to assess the country's DRM situation and corresponding 10-year strategic plan (NDCC, 2009b).
2010	Republic Act 10121 or the Disaster Risk Reduction and Management Act	— The first DRRM law in Asia (Senate of the Philippines, 2011). — It is a product of advocacy by many CSOs, academic institutions, and private groups involved in DRR and humanitarian work. — In June 2010, one month after the signature of RA 10121, then President Gloria Macapagal Arroyo approved the Executive Order 888 to adopt and implement the SNAP.

Based on NDRRMC (National Disaster Risk Reduction and Management Council). (2011). Philippine DRRM framework. Retrieved from https://www.adrc.asia/documents/dm_information/Philippines_NDRRM_Framework.pdf.

2.1 Disaster risk reduction and the Philippine development framework

The RA 10121 institutionalized the humanitarian–development nexus five years before the adoption of the 2030 Sustainable Development Agenda by the United Nations and its Member States. The law aimed to address the effects of both natural and human-induced hazards by elevating its policies and legal and institutional mechanisms, and interlinking four thematic pillars: (1) prevention and mitigation, (2) preparedness, (3) response, and (4) recovery and rehabilitation. Unlike its reactive and response-focused predecessors, the RA 10121 mandated its multilevel councils to formulate policies and plans through inclusive processes, highlighting each community's risk landscape, environment, and climate-change impacts (DRRNetPhils, 2009).

The formulation and approval of the Philippine DRRM framework prompted the completion of the first comprehensive National Disaster Risk Reduction and Management Plan for 2011–2028 (NDRRMC, 2012), building on previously identified gaps. These can be summarized as follows:
- addressing the underlying causes of vulnerability;
- linking DRR and CCA and mainstreaming these into development plans;
- tackling risk reduction to lessen future response needs;
- improving the capacities of people and institutions and providing information on DRRM and CCA; and
- focusing on building back better.

But plans are one thing; their implementation is another.

The DRRM Framework envisioned a country of "safer, adaptive, and disaster-resilient communities toward sustainable development" (NDRRMC, 2011). The law, the framework, and the national DRRM plan have facilitated the formulation of policies to improve the coherence of all multilevel and multisectoral mandated plans by mainstreaming DRR and CCA, especially in development plans (Domingo, 2016). The implementation of the law and the adoption of the plan and framework at the local level are crucial in institutionalizing DRRM (Domingo & Olaguera, 2017).

Although the linkages between the national and subnational levels have improved in the past ten years, local-government units (LGUs) continue to lag behind. Whereas local DRRM plans are compatible with the LGUs' Comprehensive Development Plans (Fig. 1.3), their interaction is not maximized due to many challenges at the local level. These include glaring gaps in terms of quality information and the lack of capacity of many

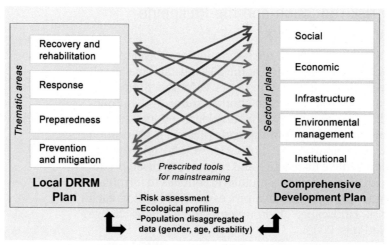

Figure 1.3 Entry points and linkages of the DRRM Plan and the Comprehensive Development Plan at the LGU level. Based on DILG (Department of Interior and Local Government). (2015). *Local planning illustrative guide: Preparing and updating the Comprehensive Development Plan.* Retrieved from https://www.dilg.gov.ph/PDF_File/reports_resources/dilg-reports-resources-2017110_298b91787e.pdf.

municipalities and cities regarding appropriate data-collection means and tools (PSA, 2018). Up-to-date national and local risk information contributes to the mainstreaming of disaster-risk and climate-change targets in local plans (Domingo & Manejar, 2018). This is also fostered by the availability of additional tools, human resources, local experts, or budgets to hire consultants or to support wider consultations and public dialogues.

Although there are mandated guidelines and tools for filling these gaps—such as the climate and disaster risk assessment and the ecological profile—LGUs' capacity to produce or source such data needs is insufficient (CLRG, 2018). Without this crucial information, the investment priorities at the community level will barely reflect the targets that are well harmonized and sufficiently elaborated at the subnational and national levels. Local funds, therefore, will tend to support programs, projects, and activities with more authority levers, including pet programs and projects of seated officials, instead of basing interventions on the results of risk assessments, as required by the RA 10121.

Moreover, oftentimes the localization and translation of governmental DRR initiatives at the community level end with the Barangay DRRM Committee and do not extend to the barangays' ordinary residents. This is also due to many factors, including poor local capacities, insufficient funding, blurry command systems, deficient management arrangements for CSO and private-sector interventions, and the lack of participatory mechanisms (Domingo & Manejar, 2018). As the receiving end of both policies and hazard impacts remains the barangay, there remains a significant gap in risk communication and capacity building, which is left for the CSOs to fill.

2.2 Challenges to a more responsive legislative amendment

The year 2015 saw the adoption of the Sendai Framework and the Sustainable Development Goals as the United Nations' response to the increasingly dynamic and complex risk ecosystem, exacerbated by the mutually reinforcing interlinkages between conflicts, food scarcity, extreme weather events, and climate change. As a result, the Philippine DRRM Plan has been recalibrated to address these global changes. Local CSOs welcomed these improvements. Even churches and faith-based institutions joined the clamor to address environmental degradation, climate change, vulnerability, and poverty, as the Vatican released in the same year the *Laudato Si'* encyclical, calling all nations to a greater paradigm shift away from a development framework that hurts the planet (Pope Francis, 2015).

Also in 2015, the RA 10121 was due for a sunset review—that is, a systematic evaluation by the Congressional Oversight Committee of its accomplishments and impacts, as well as of the organizational structure and performance of its implementing agencies. Accordingly, some groups—including the Office of Civil Defense (OCD) (Jalad, 2017), the Senate Economic Planning Office (Agub & Turingan, 2017), the Philippine Institute of Development Studies (Domingo & Olaguera, 2017), and Partners for Resilience (Partners for Resilience, 2015)—put forward their assessment of the RA 10121. In partnership with the OCD, various CSOs conducted a series of consultations in preparation for this mandated evaluation.

Although these reviews generally acknowledged the advantages of the RA 10121 regarding inclusiveness, the country suffered a series of devastating disasters between 2010 and 2015: from natural hazards such as super typhoons, floodings, and massive earthquakes, to human-induced sieges and conflicts. Mounting damage, death tolls, and internal displacements, coupled with a worsening poverty situation, put a pressure on the RA 10121's mechanisms and implementation. Bills therefore flooded both the Senate and the House of Representatives in the 17th and 18th Congresses seeking to replace the RA 10121's structure by a new centralized DRRM department.

The consolidated and approved substitute bill, the House Bill 5989, also seeks to abolish the multisectoral councils at the LGU level, which constitute the backbone of community-based DRRM. This proposal resonates in similar house bills in the Senate, including the Senate Bills 205, 331, and 1139, filed in 2019. Moreover, although the explanatory notes and policy declarations of these legislative proposals mention global frameworks, agreements, and standards, most of their provisions are highly focused on response and disaster management. CSOs, on the other hand, continue to highlight the importance of inclusive resilience building that recognizes the capacity of communities and the contributions of all sectors, thus claiming that local DRRM offices and councils should be further strengthened. These advocacy endeavors have been documented in the recently updated National DRRM Plan, which celebrates the collaboration between the government and nongovernmental institutions to

> adopt a DRRM approach that is holistic, comprehensive, integrated, and proactive in lessening the socioeconomic and environmental impacts of disasters including climate change, and promote the active involvement and participation of all sectors and all stakeholders concerned, at all levels, especially the local community (NDRRMC, 2020).

DRR indeed requires a system-dynamics lens. Linear understanding of complex risks will only result in compartmented solutions that do not address the feedback loops that spell out the creation of new risks (see Chapter 6, by Adekola and Lamond, in this volume). This is especially the case in the communities that are prone to various concomitant hazards. The current needs of the DRRM system to further reduce losses and damage, for instance, cannot be addressed by piecemeal proposals or a bloated and costly bureaucracy (Fernandez, 2020) that do not reflect the gaps acknowledged in the conducted studies and assessments. If the major identified impediments are LGUs' lack of institutional, technical, and financial capacity to formulate and implement plans (OECD, 2020), a law creating a centralizing department to replace participatory systems might not be a holistic solution. Substituting or amending the laws must ensue from comprehensive assessments, reviews, and processes that are inclusive and highly participatory, whereby problems are identified as much as progresses are acknowledged and factored in.

2.3 Method

In this chapter, we trace the evolution of the Philippine DRR legal framework as a translation of the Sendai Framework into the country's context. Our analysis draws out conclusions from presenting institutional measures supporting the Sendai Framework and identifying gaps and challenges. Then we recommend investment areas to address some of these constraints and build resilience. To concretize actions toward resilience, we present two postdisaster community projects implemented by TAO-Pilipinas (Technical Assistance Organization), a nongovernmental organization of design professionals—architects, planners, and engineers—working with poor communities in the planning, design, development, and management of settlements. TAO employs a participatory approach in its technical-assistance work, engaging directly with people's organizations and local stakeholders to facilitate and support community-driven solutions and DRR actions (Lusterio, 2018). It was involved in the rehabilitation efforts that followed two disasters triggered by extreme weather events: Tropical Storm Ketsana in 2009 and Super Typhoon Haiyan in 2013 (Fig. 1.4).

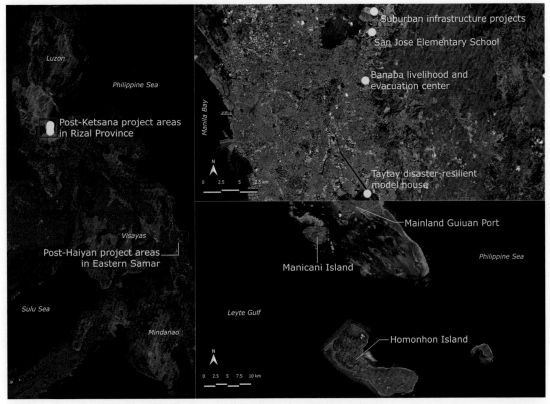

Figure 1.4 Location of TAO's post-Ketsana and post-Haiyan technical-assistance activities. Courtesy of TAO-Pilipinas, Inc. (2021).

We selected these two projects because
— they constitute responses to massive disasters that tested the technical competence and the social and management skills of the involved stakeholders: community members, TAO-Pilipinas, contractors, suppliers, social-development partners, and LGUs;
— they serve as models of an effective participatory approach to planning, design, and project implementation;
— they concern worthwhile investments toward resilience; and
— being already finalized, they are sufficiently well documented.

Our analysis looked at translations of participation, the engagement of local experts, and the use of communication tools, and how these contributed to DRR as a means to build resilience on the ground.

3. From understanding to taking action toward disaster risk reduction

3.1 Post-Ketsana response: building disaster-resilient community infrastructures

When Tropical Storm Ketsana hit the national capital region in September 2009, its accompanying rains submerged vast areas of Metro Manila and the neighboring Rizal province. In what seemed to be a relatively mild Category-1 storm, the unusually high volume of rain (450 mm in a 12-hour period) brought the entire metropolis to a standstill and its residents experienced the most extensive flooding in recent memory. People living in low-lying areas were caught unaware as muddy and garbage-strewn floodwaters clogged drainage systems and inundated their homes. Close to five million people were affected by the 100-year flood, and the estimated damage on infrastructures was placed at PHP4.3 billion (around USD90 million), accounting for 185,000 totally and partially damaged houses (NDCC, 2009a).

Widespread flooding not only damaged houses but also affected power- and water-utility services, disrupting economic and social activities. Urban planners attributed the flooding to a combination of factors (von Einsiedel, 2010). They contended that, at the regional level, it was caused by the denudation of forest and watershed areas in the Sierra Madre mountain range, the urbanization of the Marikina floodplains, rising sea levels affecting coastal and low-lying land, the silting of rivers, and the obstruction of waterways. Moreover, inadequate and substandard infrastructures, uncontrolled urban growth, and weak urban management, including as regards solid wastes, exacerbated the flooding at the local level. With the devastation concentrated in the nation's capital, Ketsana somehow elevated the public discourse on climate-change and urban-disaster resilience.

Ketsana severely affected the urban poor living in informal settlements situated along waterway easements and within floodway embankments, housed in light-material nonengineered structures. Tens of thousands sought shelter in crowded evacuation centers and tried to save what was left of their damaged houses. In the aftermath of Ketsana, amid lives, properties, and livelihoods lost, the urban poor were inordinately blamed for the flooding because their houses obstructed the waterways. Moreover, the urban poor were thrice victimized (Balderrama, 2012), as they also faced threats of eviction associated with the government's typical postdisaster response: off-city relocation of informal settlers that distances them from their social connections, existing sources of livelihood, and access to urban services and opportunities.

The international relief and development agency Christian Aid provided rehabilitation assistance to the most affected and vulnerable communities in the Rizal province. It supported local NGOs working in four communities: Barangay Banaba in San Mateo; Sabah and Suburban in San Jose, Rodriguez; and Sitio Lumang Ilog in San Juan, Taytay. Christian Aid's Ketsana rehabilitation program set out to rebuild lives and restore livelihoods, enhance community resilience to disasters, and reduce vulnerabilities. Seven NGOs—a mix of area-based implementers and resource organizations, including TAO-Pilipinas—worked together as a consortium. The consortium implemented key interventions that included developing community-managed livelihood enterprises, setting up small DRR community-infrastructure facilities, implementing technical capacity building, and conducting geohazard and flood-risk assessments. It also sought to strengthen DRR capacities of people's organizations and advocacy work toward DRR policies that recognize the right to the city of the urban poor living in hazard-prone areas.

TAO-Pilipinas implemented the small-infrastructure component of the Ketsana rehabilitation program from 2011 to 2013. As the technical resource NGO partner, it was in charge of developing community design and planning capacities on disaster- and climate-resilient housing and small buildings, and building small infrastructures that would contribute to the overall resilience of the targeted communities. TAO led the design and construction (or upgrading) of several greenfield or preexisting structures: an evacuation and livelihood center in Banaba; a model disaster-resilient house in Taytay; a multipurpose center, an evacuation facility (covered court), and a pedestrian bridge in Suburban; and the drainage line in Sabah's evacuation facility (San Jose Elementary School). The largest of these facilities was the Banaba Evacuation and Livelihood Center, a three-story structure built on an 827-m^2 lot to house up to 89 evacuee families during an emergency (Fig. 1.5) (Varona, 2012). This is aligned with the provision of the Department of Interior and Local Government's Memorandum Circular 122 of 2018, which directed all

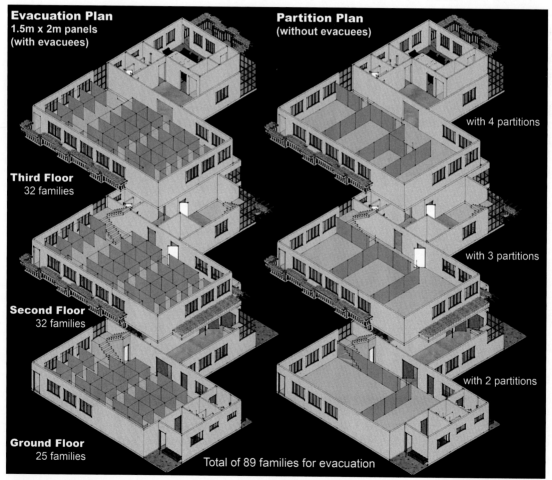

Figure 1.5 Interior configurations of the Banaba Evacuation and Livelihood Center. Courtesy of TAO-Pilipinas, Inc. (2012).

LGUs to build an evacuation center according to set guidelines (DILG, 2018a, 2018b).

The project consortium saw the application of disaster-resilient principles in building these facilities as a long-term strategy for disaster preparedness. Building-code standards on easements and space allocation, geotechnical studies, and wind and seismic-load considerations guided the design and construction of the infrastructures. In the Taytay model house, for instance, the liquefaction-prone site was a critical factor in the design of the structure's foundation. The model house thus demonstrated the mat-type foundation as an engineering intervention appropriate for the area's soil condition.

Integral to TAO's assistance was a technical capacity-building program intended to increase community knowledge on DRR regarding site planning as well as house design and construction. In the span of three months, about 30 representatives of the target communities underwent four training sessions on the technical aspects of developing disaster-resilient sites and structures. The training topics covered planning considerations and design standards for housing developments, basic comprehension of technical plans and documents, building materials and technologies for low-cost housing, and engineering principles for building and infrastructure design. TAO also presented to the communities the idea of retrofitting existing houses as a preventive measure against typhoon and earthquake damage.

Hands-on learning was an important part of community training. TAO developed the workshop toolkits and simulated site-planning and design exercises to apply what participants learned from the lecture inputs. TAO selected Suburban Phase 1B, a government housing site for relocated informal settlers, as a demonstration area for house retrofitting. Suburban is prone to both earthquakes and floods since parts of the site are low lying and situated near a seismic fault trace and a creek. Because the settlers were already beneficiaries of a governmental housing program, relocation to a safer site was not deemed an option, even if they would keep exposed to hazards. The community thus had to learn how to cope with the risk of living near fault lines and in substandard housing units by becoming more aware of earthquakes, planning for disaster mitigation, and taking measures (like self-built improvements) to reduce their vulnerability.

Onsite, the training participants learned to use a visual-assessment tool for inspecting and identifying construction deficiencies in ten houses. Local structural-engineering consultants provided technical advice on appropriate retrofitting measures. TAO organized the participants and other community

Figure 1.6 A structural engineer instructs community builders during the hands-on demonstration of house retrofitting. Courtesy of TAO-Pilipinas, Inc. (2012).

members to execute minor retrofitting works in three houses in Suburban under the engineers' direction (Fig. 1.6). The retrofitted buildings served as demonstration cases to guide the self-help retrofitting of houses in the community.

TAO's intervention took on a highly participatory approach, with the communities themselves determining what small infrastructures needed to be built or improved through the Ketsana rehabilitation program. The communities identified the facilities that would provide them with safe-evacuation structures and temporary refuge during strong typhoons and other hazardous events. Likewise, the planning, design, and construction of these facilities called for the active engagement of people's organizations in the communities. Whenever possible, TAO made room for the community's direct involvement in project implementation for them to value it as their own. The buildings' designs derived from participatory planning and design workshops. The consortium resorted to community contracting to implement the constructions, thus providing income and skills' improvement for local workers and also an opportunity to transfer technical knowledge on disaster-resilient construction. Yet, this approach experienced some difficulties in terms of construction management. Engaging a community-based workforce with inadequate skills and equipment affected workmanship and slowed the projects. Before the facilities' completion, TAO also discussed postconstruction responsibilities with the communities, including regarding the management of the structures.

To this end, TAO initiated a workshop on the formulation of community guidelines for the use and maintenance of the small infrastructures for DRR.

Besides the development of these infrastructures, TAO-Pilipinas complemented the work of other NGO partners in the consortium, namely Community Organizers Multiversity (COM) and Partnership of Philippine Support Service Agencies (PHILSSA). With COM, TAO built the capacities of community leaders on subdivision planning and design. TAO collaborated with PHILSSA in conducting a participatory research on the situation of the communities regarded by government agencies as living in "danger areas." TAO provided technical inputs on topography, geophysical features, and land-use and development plans of local governments, and prepared data-gathering tools for local research teams. Community leaders participated in the data analysis, deepening their awareness of disaster risks in their communities. TAO also convened a series of roundtable discussions on government flood-control projects and plans to push for the urban poor's participation in DRR-related technical studies and development planning.

TAO's Ketsana response shows the considerable extent of participation in building infrastructures for DRR—in planning, design, and construction—and that these create capacity-building opportunities for community resilience. The key to meaningful involvement was the development of technical-knowledge capacities at the people's organization level. Technical-knowledge transfer, incorporated into planning consultations and design workshops, allowed communities to make informed decisions that influenced the project implementation. The program also showed the potential of an effective system of community contracting as a venue for resilience building. The project implementers had to address community-contracting problems related to inadequate resources, namely financial capital, equipment, suppliers, and construction workers. They had to target investments in the technical training of community builders because local carpenters, masons, and other construction workers were the ones likely to engage in self-built house construction in urban poor settlements.

3.2 Post-Haiyan response: participatory planning, design, and construction

On November 8, 2013, the Philippines bore the brunt of what was then the most powerful typhoon in recorded history. In a country that is a hotspot for the world's strongest tropical cyclones, whatever DRR measures put in place proved no match to the 275-km-per-hour winds packed by the landfall of Super Typhoon Haiyan. Haiyan swept across the eastern, central, and western Visayas regions, triggering storm surges and flash floods, with some communities completely washed out. It also destroyed power lines, water supply, and communication lines. In Guiuan, Eastern Samar, where Haiyan made its first landfall, almost everything was damaged if not flattened—infrastructures, coconut plantations, houses, and boats. Haiyan left in its wake catastrophic destructions, with 6300 fatalities and PHP95 billion (around USD2.2 billion) in total damage (NDRRMC, 2013).

Two months after Haiyan, TAO-Pilipinas initiated a shelter-damage survey in the remote islands of Manicani and Homonhon in Guiuan. This initiative was part of a postdisaster needs and capacities assessment organized by PMPI (Partnership Mission for People's Initiative, formerly Philippine Misereor Partnership Inc.). Located at a 45-minute motorized-boat ride from the port in mainland Guiuan, Manicani is a small island (1165 ha) with about 2300 residents living in four barangays. A 20-km long island with about 4200 residents in eight barangays, Homonhon is at a three- to four-hour boat ride away from the Guiuan port. Rich in mineral resources, these islands have government-permitted open-pit mines despite being proclaimed by the Department of Environment and Natural Resources as protected areas. PMPI, a network of NGOs and faith-based people's organizations, opposes these destructive mining operations and advocates for sustainable agriculture and coastal-resource management in the islands (PMPI, 2019).

PMPI led the disaster-response intervention dubbed Project Pagbangon, a three-year program aimed at rehabilitating Manicani and Homonhon into sustainable island ecosystems. PMPI formed a consortium of NGO partners to implement the project's seven components: coastal-resource management; health; shelter and evacuation; renewable-energy, communication, and water infrastructures; land-resource management and sustainable agriculture; community-based DRRM; and capacity building in community organizations.

TAO-Pilipinas implemented the program's shelter and evacuation-center component under the objective of reducing vulnerability to geohazards and managing disaster and climate risks. PMPI tasked TAO to deliver 120 disaster-resilient permanent homes for affected families and construct 12 barangay evacuation centers in the islands. The cornerstone of TAO's intervention in Project Pagbangon remained the participatory approach, applied to the extent feasible in all aspects of its implementation—from the selection of the shelter beneficiaries and sites, to planning, design, and construction of the houses and evacuation centers.

The participatory approach enabled the project's beneficiaries to significantly influence the outcomes of the planning and design process. Such an approach necessitates knowledge transfer and capacity building for them to be aware of their options and limitations to make informed decisions. TAO facilitated these through a series of participatory workshops and consultations that oriented beneficiaries and stakeholders about sustainable-settlement planning and disaster-resilient design and construction.

Mindful of geohazard considerations and adhering to the "building back better" principle, TAO worked with geologists to conduct hazard assessments and site-suitability studies for the proposed locations of the houses and evacuation centers. The assessed hazards included typhoons, storm surges, floods, landslides, tsunamis, and liquefactions. This informed TAO and the beneficiaries where the safe locations were (that is, the areas with low combined geohazard susceptibility) and where to adopt structural design modifications for mitigation (in the absence of low-risk areas) (Lusterio, Sarraga, & Matabang, 2019).

Project beneficiaries also underwent a participatory house-design workshop (Fig. 1.7), to generate schemes for a typical disaster-resilient permanent shelter. Incorporating people's articulated needs and preferences and considering budgetary

Figure 1.7 Participatory design workshop with Manicani shelter beneficiaries. Courtesy of TAO-Pilipinas, Inc. (2015).

Figure 1.8 Typical houses built in Manicani (left) and Homonhon (right). Courtesy of TAO-Pilipinas, Inc. (2015, 2017).

constraints, TAO developed the final design of the houses to structurally withstand 220-km-per-hour winds (Fig. 1.8). In Manicani, this took the form of 25-m² houses with a regularly shaped compact plan, a hip roof at 30-degree pitch with trussed-roof framing and narrow eaves, windows with shutters, gutters for rainwater harvesting, and a concrete structure with adequate steel reinforcement and connections. In Homonhon, the design features were almost the same except for the use of microconcrete roof tiles instead of galvanized-sheet roofing. Houses built in moderate- to high-risk areas had a modified design, elevated above the natural ground (on fill or on concrete pedestals).

One of TAO's biggest implementation challenges was the participatory shelter construction, especially in such remote island settings. Participation was facilitated by having the beneficiaries select local contractors to construct their houses. The beneficiaries also contributed with labor and materials, helping in the hauling and safeguarding of materials, and monitoring the progress of local contractors' work. The project experienced logistical difficulties in bringing construction materials to the islands and delays due to inclement weather. Cultural norms also caused project setbacks, from the generally laid-back attitude of islanders to some local contractors abandoning construction work to celebrate a barangay fiesta. Nonetheless, participation managed overall to instill ownership and pride for the project outcomes. Local contracting generated temporary employment for residents and also facilitated technical-knowledge transfer on disaster-resilient construction methods (Lusterio et al., 2019).

Before construction completion, TAO held another workshop to guide beneficiaries in the use, maintenance, repair, and future expansion of the structures. This made beneficiaries aware about how to increment their houses without compromising their structural integrity and disaster-resilient features. Posters and pamphlets featuring these lessons were also disseminated so that the beneficiaries could apply them when TAO is no longer present.

TAO employed a similar participatory planning, design, and construction process in building the barangay evacuation centers. Project Pagbangon coordinated with local authorities to find suitable land and secure permits to build these. The barangay LGUs approved the construction project through council resolutions. The project set up evacuation-center committees, involving the local population in construction monitoring and, subsequently, in the facilities' operation and maintenance. Each 100-m^2 evacuation center can accommodate about 100 persons during an emergency. These structures were also equipped with solar-lighting and rainwater-harvesting systems.

The communities involved in Project Pagbangon were not just passive beneficiaries. They were key decision-makers, informed and empowered to lay their own paths toward resilience. With strong community engagement, the project delivered 118 shelter units and 11 evacuation centers in the islands. After the project's completion, these disaster-resilient houses have already served as alternative evacuation areas in the community, accommodating neighbors and relatives without safe shelter during typhoons (PMPI, 2019). Most beneficiaries have also continued to invest in their houses, building additional rooms and kitchen extensions with the available funds (Lusterio et al., 2019). With sturdy homes located in less susceptible areas of the island and evacuation centers in place for temporary refuge during hazardous events, the project gave a renewed sense of hope and safety to communities nearly leveled by one of the most devastating disasters to hit the country.

3.3 On effective participation

The two cases illustrate the local technical competence of a multidisciplinary group of professionals engaged in DRR. Local actors tend to go straight to the root of the problems because they know the related context and culture, and speak the concerned people's language. Participation came in different levels with the accompanying capacity building and accountability. Participation is not just attending meetings. It is for people to speak their hearts out and make decisions that shape the final form of the interventions. Participation is consciously making decisions that affect their lives.

Participation at the community level requires building consensus and an impartial assessment of the issues at hand, such as selecting the beneficiaries for house construction or retrofitting when the need is higher than what a project can provide. Conversely, participation at the household level requires decisions guided by personal choices, such as where and how to build a house. In the gender-biased Philippine society, final decisions are made by the husband, but the wife actively participates in the consultations because these are nonearning activities. When the husband disagrees with the choices or decisions made by the wife during the consultations, community consensus during implementation is affected. Hence, decisions concerning shelter require agreement at the household level too.

4. Practices, structures, and linkages: how disaster risk reduction interventions can last

4.1 Practices: what people learned, retained, valued, and applied

The complexity of participation and decision-making determines the preparations for knowledge transfer in DRR interventions. Communication tools and teaching methods that help laypeople grasp technical information are crucial. Hands-on learning exercises and 3D models facilitate better understanding and visualization of buildings' architectural design and structural assessment. In the two postdisaster projects presented, laypeople were transformed into *para-architects/engineers* after two to three days of effective learning and gained knowledge that they have applied in design workshops, structural assessments, actual constructions, and later on in their houses' maintenance and extension. The capacity building of house owners and local builders in sustainable design and disaster-resilient construction guided the translation of drawings and DRR legislation into buildings. The two cases showed how effectively ordinary residents in both rural and urban contexts are capable of technical learning, analyses, and design, as well as of applying these and making decisions. This is how we define and translate community participation.

4.2 Structures: understanding sustainable and disaster-resilient design and construction

Investing in DRR includes committing long-term resources for structural measures that can withstand hazards and last. This requires designing for the site's specific conditions, rather than applying ready-made solutions coming from a different context. In both post-Ketsana and post-Haiyan recovery processes, building sustainable and disaster-resilient houses and evacuation centers was guided by the Philippine design standards, enforced from the plan to the construction site. However, laypeople often did not understand the prescription of design standards applied to the houses they live in. The hands-on training gave the building owners confidence over the structural integrity of their houses and knowledge on how they could safely expand these to accommodate their families' changing spatial needs. This means cocreation and empowerment, and characterizes TAO-Pilipinas' technical interventions: capability building and direct engagement of affected populations with practical application of the knowledge gained. This is not sweat equity, a labor contribution done with or without understanding the whole construction process.

However, this perspective generally contrasts with conventional postdisaster responses. For instance, in the post-Haiyan experience, the Shelter Cluster set guidelines for disaster-resilient construction that were interpreted in various ways (Opdyke, Javernick-Will, & Koschmann, 2017). Super Typhoon Haiyan thus gave birth to many shelter models, most of which were for emergency or transitional use with a short lifespan, failing to deliver more sustainable solutions.

A disaster-resilient shelter should move toward permanent housing. But obstacles such as land tenure, lack of livelihood, and geohazards often deter the quick implementation of lasting postdisaster housing solutions. These barriers should be considered and addressed through proactive planning at the LGU level. Self-built houses often do not conform to construction guidelines or standards and may end up creating disaster risks instead of reducing them. Investment must also be made on the capacity building of local builders to ensure that they fully understand disaster-resilient construction and apply its principles in their work. Besides ensuring safe housing, governmental agencies should care for the provision of evacuation centers, which represent capital-intensive commitments that are worthwhile investments to protect life.

4.3 Linkages: civil-society organizations' advocacies and the consortium approach

DRR is not a one-person performance, but a concerted effort of all stakeholders touching on development's social, economic, and environmental dimensions. It entails forming communities, for example, into people's organizations, farmer cooperatives, and fisherfolk associations, so that they can speak with one voice and join networks to be recognized, access resources, and contribute to development and in the discourses affecting them. This effort requires economic and livelihood-development experts to help uplift the poor from poverty. It also requires experts in both built and natural environments to enlighten all of creation about the value of the Earth and how it cradles life (Pope Francis, 2015).

Finding order and defining structures, recognizing and respecting roles and contributions, engaging the affected population, and applying a bottom-up approach in decision-making remain guiding principles for effective multistakeholder DRR interventions. Such interventions should translate into a proactive approach with a focus on prevention and mitigation, as well as adaptation and preparedness—rather than only on relief and response. They should strengthen engagement at the local levels, from the municipality or the city to the barangays and the communities, who are the first responders in disasters. The Philippine CSOs will strongly resist the government efforts to undermine that window for participation, as shown in the many proposed bills to discard the RA 10121. The institutional response to DRR has been a struggle between the centralized and participatory approaches for more than half a century in the Philippines. So it is expected that the tug of war will continue.

The consortium approach has stood out as a localized response to massive postdisaster situations. No one organization has all the needed expertise, whereas a multidisciplinary pool of local experts can work together to comprehensively face such a demanding context. The strength and weakness of a consortium lie in its coordination. The success of the Project Pagbangon consortium ensued from the very strong secretariat coordinating the complex activities across the partners, as well as from the partners' openness to collaborate and cooperate rather than compete. The consortium approach could also be applied to city-wide or regional development projects. The related interventions can expand from the community to the city or municipality. Whereas engagement at the national level may not be fully favorable to a truly participatory approach, engaging with the city or municipality can result in a strongly localized collaboration between CSOs and the LGUs. Of course, the government must remain accountable and CSOs should continue to remind the officials about this.

5. Toward stronger community participation and localization for smarter disaster risk reduction investment

Regardless of the Sendai Framework, investing in DRR should put people at the center of development, whether it is for post-disaster response or for long-term sustainability. Disaster brings in urgency to action, but sustainability must be the long-term goal. Empowered communities can make informed decisions that move them away from risks. The challenge is how to institutionalize a DRR stance that remains people-centered. Just as we need an enlightened citizenry, we also need an enlightened bureaucracy. The current institutional framework for DRR espoused by the RA 10121 calls for participation that cuts across to the lowest level of governance, emphasizing the role of LGUs and DRR councils in disaster-response decision-making and implementation. It also calls for a proactive approach through strengthened prevention and preparedness measures infused with CCA alongside response and relief.

As the urban poor, farmers, and fisherfolk often occupy susceptible areas, their vulnerability is exacerbated by the concomitant exposure to hazards and poverty. Investing in poverty reduction, organizational development, and the strengthening of these sectors, especially in rural areas, would increase their ability to demand space for participation, be heard, and influence development and intervention decisions at the local and even national levels. This is a long journey that must go beyond disaster response.

At the same time, the inevitable onslaught of hazards requires investments in structural measures. Investing in infrastructures such as permanent shelters and evacuation centers is more a necessity than a choice. Shelter response should consider sturdy houses that could gradually become permanent. The humanitarian emergency-shelter response often provides for immediate relief, but its short-term efficacy needs some reflection. The response to massive postdisaster shelter needs will continue until a sustainable solution emerges from proactive measures of institutionalizing and mainstreaming DRR in national and local development and investment plans. Governments should proactively engage communities in shelter planning and development, complemented with socioeconomic development programs encouraging voluntary resettlement and the creation of sustained communities in safe places. This would avoid a chaos of displaced people uncertain of their future, as experienced after Haiyan (Thomas, 2015).

In planning settlements, the geologic and meteorological hazard mapping cannot be discounted. This is critical for building better resettlement sites as well as on-site development options. In urban areas, on-site development options for informal settlements constitute a big challenge with their massive scale. Where self-built unengineered houses dominate urban informal settlements, the risks are high due to vulnerabilities coupled with human-made hazards, specifically the poor quality of the built fabric, which is prone to fire or to collapse during earthquakes. When former informal dwellers were resettled to a flood- and earthquake-prone area in Rodriguez, the risks due to exposure to natural hazards were high. To reduce disaster risk in settings where exposure is unavoidable, a major investment in structural measures is needed. Measures such as site upgrading, house retrofitting, or provision of legal access to utilities help improve the physical condition of the settlement, reduce vulnerability, and increase resilience. These are major areas for governmental investments.

So investing in DRR for resilience starts with putting people at the center of development. This includes investing in people's capacity building and sturdy infrastructures that conform to local design standards. It also comprises bonding partners together as a consortium and relying on local experts, to face the bigger challenges heralded by the climate-change prospects. DRR investments should also advocate against greenhouse-gas-emitting projects, a major challenge given the power of money. But we need to continue persisting even if the task looks insurmountable.

References

ABS-CBN (Alto Broadcasting System — Chronicle Broadcasting Network). (2019). *'Quake alley': Philippines tallies 'normal' average of 20 quakes a day—PHIVOLCS*. Retrieved from https://news.abs-cbn.com/video/news/04/25/19/quake-alley-philippines-tallies-normal-average-of-20-quakes-a-day-phivolcs.

Agub, S. B., & Turingan, P. A. S. (2017). *Examining the Philippines' disaster risk reduction and management system*. SEPO Policy Brief 17-01. Retrieved from http://legacy.senate.gov.ph/publications/SEPO/PB_Examining%20PH%20DRRM%20System_Revised_27June2017.pdf.

Balderrama, B. (2012). Standing up for secure and safe settlements: Networking and policy advocacy by Ketsana-affected communities. In A. M. A. Karaos (Ed.), *Resilient urban communities: Stories from the Ketsana Rehabilitation Programme* (pp. 98–109). Quezon City: Christian Aid.

Benson, C. (2009). *Mainstreaming disaster risk reduction into development: Challenges and experience in the Philippines*. Retrieved from http://www.unisdr.org/preventionweb/files/8700_8700mainstreamingphilippines1.pdf.

Boquet, Y. (2017). *The Philippine Archipelago*. Cham: Springer.

CLRG (Center for Local and Regional Governance). (2018). *IRM Post-conference. Policy Brief*. Retrieved from https://localgov.up.edu.ph/uploads/1/4/0/0/14001967/irm_policy_brief.pdf.

COA (Commission on Audit). (2014). *Evolution of disaster laws in the Philippines*. Retrieved from https://www.coa.gov.ph/disaster_audit/article2.html.

DILG (Department of Interior and Local Government). (2018a). *DILG to LGUs: Use local DRRM fund to build evacuation centers*. DILG News. Retrieved from https://dilg.gov.ph/news/DILG-to-LGUs-Use-local-DRRM-fund-to-build-evacuation-centers/NC-2018-1243.

DILG (Department of Interior and Local Government). (2018b). *Guidelines for local government units on the strengthening of evacuation systems using the Local Disaster Risk Reduction and Management Fund (LDRRMF)*. Memorandum Circular no. 122 series of 2018. Retrieved from https://dilg.gov.ph/issuances/mc/Guidelines-for-Local-Government-Units-on-the-Strengthening-of-Evacuation-Systems-using-the-Local-Disaster-Risk-Reduction-and-Management-Fund-LDRRMF/2773.

Domingo, S. N. (2016). *An assessment of the sectoral and institutional implementation of the NDRRMP*. Philippine Institute for Development Studies Discussion Paper Series No. 2016-49. Retrieved from https://pidswebs.pids.gov.ph/CDN/PUBLICATIONS/pidsdps1649.pdf.

Domingo, S. N., & Manejar, A. J. A. (2018). *Disaster preparedness and local governance in the Philippines*. Philippine Institute of Development Studies Discussion Paper Series No. 2018-52. Retrieved from https://pidswebs.pids.gov.ph/CDN/PUBLICATIONS/pidsdps1852.pdf.

Domingo, S. N., & Olaguera, M. D. C. (2017). *Have we institutionalized DRRM in the Philippines?*. Philippine Institute of Development Studies Policy Notes No. 2017-12. Retrieved from https://pidswebs.pids.gov.ph/CDN/PUBLICATIONS/pidspn1712.pdf.

DRRNetPhils (Disaster Risk Reduction Network Philippines). (2009). *DRRM Bill*. Retrieved from https://www.preventionweb.net/files/11448_PDCdrrmbill primer.pdf.

Eckstein, D., Künzel, V., Schäfer, L., & Winges, M. (2019). *Global Climate Risk Index 2020*. Germanwatch briefing paper. Retrieved from https://www.germanwatch.org/sites/germanwatch.org/files/20-2-01e%20Global%20Climate%20Risk%20Index%202020_14.pdf.

EMI (Earthquakes and Megacities Initiative). (2007). *Disaster Risk Management Master Plan of Metro Manila*. EMI Brochure BR-07-02. Retrieved from https://www.preventionweb.net/files/1499_3cd2007DRMMPMMBR0702.pdf.

Fernandez, B. (2020). *Philippines: New disaster department costly, needless—Senators*. Retrieved from https://businessmirror.com.ph/2020/11/02/new-disaster-department-costly-needless-senators.

Government of the Philippines. (2020). *Coastal communities*. Retrieved from https://www.gov.ph/web/green-climate-fund/coastalcommunities.

Government of the Philippines. (2021). *About the Philippines*. Retrieved from https://www.gov.ph/es/the-philippines.html.

Jalad, R. B. (2017). *Review of Republic Act 10121 and highlights of proposed amendments*. Retrieved from https://static1.squarespace.com/static/58ec1df1d1758e3915cb1470/t/59017b27d482e987e3ac92ad/1493269362772/Jalad-Purisima+-+NDRRMC+-+Review+of+RA+10121.pdf.

Llosa, S., & Zodrow, I. (2011). *Disaster risk reduction legislation as a basis for effective adaptation*. Retrieved from https://www.preventionweb.net/english/hyogo/gar/2011/en/bgdocs/Llosa_&_Zodrow_2011.pdf.

Lusterio, A. (2018). The power of the people's plan. In L. Mendoza-Rivera (Ed.), *TAO-Pilipinas architects in the margins* (pp. 1−12). Quezon City: TAO-Pilipinas.

Lusterio, A., Sarraga, V., & Matabang, G. (2019). *Battling waves: Lessons in humanitarian shelter response in the small islands of Manicani and Homonhon, Eastern Samar and geographically isolated and disadvantaged areas in Northern Samar*. Quezon City: TAO-Pilipinas.

NDCC (National Disaster Coordinating Council). (2009a). *NDCC update: Final report on Tropical Storm "Ondoy" (Ketsana) and Typhoon "Pepeng" (Parma)*. Retrieved from https://ndrrmc.gov.ph/attachments/article/1543/Update_Final_Report_TS_Ondoy_and_Pepeng_24-27SEP2009and30SEP-20OCT2009.pdf.

NDCC (National Disaster Coordinating Council). (2009b). *Strengthening disaster risk reduction in the Philippines: Strategic National Action Plan (2009−2019)*. Retrieved from https://www.adrc.asia/countryreport/PHL/2009/PHL_attachment.pdf.

NDRRMC (National Disaster Risk Reduction and Management Council). (2011). *Philippine DRRM Framework*. Retrieved from https://www.adrc.asia/documents/dm_information/Philippines_NDRRM_Framework.pdf.

NDRRMC (National Disaster Risk Reduction and Management Council). (2012). *National Disaster Risk Reduction and Management Plan (NDRRMP) for 2011−2028*. Retrieved from https://ndrrmc.gov.ph/attachments/article/41/NDRRM_Plan_2011-2028.pdf.

NDRRMC (National Disaster Risk Reduction and Management Council). (2013). *NDRRMC update: Final report on the effects of Typhoon "Yolanda" (Haiyan)*. Retrieved from https://ndrrmc.gov.ph/attachments/article/1329/FINAL_REPORT_re_Effects_of_Typhoon_YOLANDA_(HAIYAN)_06-09NOV2013.pdf.

NDRRMC (National Disaster Risk Reduction and Management Council). (2020). *National Disaster Risk Reduction and Management Plan (NDRRMP) 2020−2030*. Retrieved from https://ndrrmc.gov.ph/attachments/article/4147/NDRRMP-Pre-Publication-Copy-v2.pdf.

OECD (Organisation for Economic Co-operation and Development). (2020). *Common ground between the Paris Agreement and the Sendai Framework: Climate change adaptation and disaster risk reduction*. Retrieved from https://doi.org/10.1787/3edc8d09-en.

Opdyke, A., Javernick-Will, A., & Koschmann, M. (2017). *Typhoon Haiyan shelter case studies*. Retrieved from https://www.colorado.edu/lab/gpo/sites/default/files/attached-files/opdyke_et_al_2017_typhoon_haiyan-shelter_case_studies.pdf.

Partners for Resilience. (2015). *Before sunset—PFR inputs to RA 10121 Sunset Review*. Retrieved from https://resilientphilippines.com/wp-content/uploads/publications/Before%20Sunset%20-%20PFR%20Inputs%20to%20RA10121%20sunset%20review.pdf.

PHIVOLCS (Philippine Institute of Volcanology and Seismology). (2020). *Destructive earthquakes in the Philippines*. Retrieved from https://www. phivolcs.dost.gov.ph/index.php/earthquake/destructive-earthquake-of-the-philippines.

PHIVOLCS (Philippine Institute of Volcanology and Seismology). (2021). *Earthquake information*. Retrieved from https://www.phivolcs.dost.gov.ph/index.php/earthquake/earthquake-information3.

PMPI (Partnership Mission for People's Initiative). (2019). *Project Pagbangon: Stories of resilience, partnership and change*. Quezon City: PMPI.

Pope Francis. (2015). *Laudato Si': On care for our common home* [Encyclical]. Retrieved from http://www.vatican.va/content/francesco/en/encyclicals/documents/papa-rancesco_20150524_enciclica-laudato-si.html.

PSA (Philippine Statistics Authority). (2017). *2015 Census of population. Report no. 2—Demographic and socioeconomic characteristics*. Retrieved from https://psa.gov.ph/sites/default/files/2015 CPH_REPORT NO. 2_PHILIPPINES.pdf.

PSA (Philippine Statistics Authority). (2018). *Philippine Statistical Development Program 2018*. Retrieved from https://psa.gov.ph/philippine-statistical-system/psdp/chapters.

PSA (Philippine Statistics Authority). (2019). *Proportion of poor Filipinos was estimated at 16.6 percent in 2018*. Retrieved from https://psa.gov.ph/poverty-press-releases.

PSA (Philippine Statistics Authority). (2020). *Farmers, fisherfolks, individuals residing in rural areas and children posted the highest poverty incidences among the basic sectors in 2018*. Retrieved from https://psa.gov.ph/poverty-press-releases/nid/162541.

Senate of the Philippines. (2011). *Guingona to speak about Philippines' DRRM Law (RA 10121) before other Asian countries*. Press release. Retrieved from http://legacy.senate.gov.ph/press_release/2011/0905_guingona1.asp.

Thomas, A. R. (2015). *Resettlement in the wake of Typhoon Haiyan in the Philippines: A strategy to mitigate or a risky strategy?* Retrieved from https://www.brookings.edu/wp-content/uploads/2016/06/Brookings-Planned-Relocations-Case-StudyAlice-Thomas-Philippines-case-study-June-2015.pdf.

Varona, M. F. (2012). Building resilience through self-help infrastructure development and technical capacity-building. In A. M. A. Karaos (Ed.), *Resilient urban communities: Stories from the Ketsana Rehabilitation Programme* (pp. 88—97). Quezon City: Christian Aid.

von Einsiedel, N. (2010, June). What went wrong? The root of widespread flooding in Metro Manila and what we can do about it. *TAO Shelter Magazine, 7—8*, 7—9. Retrieved from https://tao-pilipinas.org.

Investing in the social places of heritage towns

Tomoko Kano[1], Takae Tanaka[2] and Momoyo Gota[3]
[1]*Teikyo Heisei University, Tokyo, Japan;* [2]*Tama University, Kanagawa, Japan;*
[3]*Tokyo University of Science, Tokyo, Japan*

1. The importance of social places of heritage towns in disaster risk reduction

Worldwide, heritage sites and artifacts have survived various disasters triggered by natural and human-driven hazards. The memories of the destruction of local properties and places caused by the 2011 Great East Japan Earthquake are still fresh in the minds of Japanese people. However, several social places, such as temples, shrines, and other heritage sites, withstood this massive disaster and housed many earthquake and tsunami survivors, including tourists. Following the disaster, locals and tourists stayed together inside the shrines for several days, as they provided safe environments. In this manner, cultural heritage sites not only survive disasters but also can ensure people's safety. According to the Sendai Framework for Disaster Risk Reduction 2015—2030, disaster recovery, rehabilitation, and reconstruction phases, which should be planned before the occurrence of a disaster, provide a critical opportunity to "Build Back Better." They enable the integration of disaster risk reduction into development measures and, thereby, support making nations and communities resilient.

This study advocates a method to rebuild resilient communities, as indicated by the third priority of the Sendai Framework. Furthermore, it investigates the potential of social places of heritage towns in reducing disaster risks and enhancing community resilience. Through this research, we promoted the collaboration between architects and anthropologists to examine risk-mitigation measures and disaster-recovery processes followed in World Heritage towns, which consist of historical buildings (including dwellings), in the Asiatic seismic zone.

Investing in Disaster Risk Reduction for Resilience. https://doi.org/10.1016/B978-0-12-818639-8.00009-0

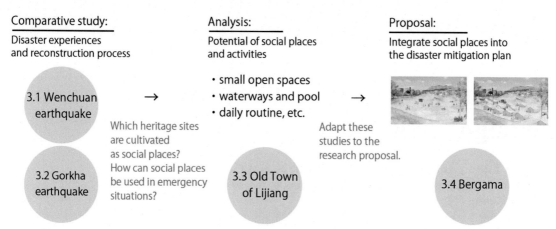

Figure 2.1 The research process followed in Sichuan, Kathmandu, Lijiang, and Bergama. The authors.

2. Methodology for studying social places in different heritage towns

In this research, we first conducted a comparative study of two earthquakes, Wenchuan (China) and Gorkha (Nepal), and collected data on the local disaster experiences and recovery processes (Fig. 2.1). Subsequently, we discussed the potential of social places in facilitating disaster preparedness in the Old Town of Lijiang (China), by addressing the following questions: Which heritage sites are cultivated as social places? How can social places and activities be used in emergency situations? Through these analyses, we proposed a disaster risk mitigation plan that involves social places and activities and integrates community-based disaster risk reduction measures for the World Heritage town of Bergama (Turkey). We collected data from previous studies, as well as from fieldwork, which included observations, interviews, and a spatial survey, as indicated in Table 2.1.

In addition, we held an international professional meeting and a community-based workshop in Bergama in February 2019, to increase local people's awareness of disaster risk preparedness. This can actively support the development of community-based disaster risk management approaches throughout the tourism industry, which is the region's key economic driver.

3. Social places of different heritage towns

3.1 The reconstruction process following the Wenchuan Earthquake

The Wenchuan Earthquake occurred on May 12, 2008 in the Sichuan Province, China. This 8.0-magnitude quake caused

Table 2.1 Research steps followed in Sichuan, Kathmandu, Lijiang, and Bergama.

Site	Date	Method
Sichuan, China	September 13–17, 2016	Observation of exhibitions on earthquake ruins and museums in Beichuan County. Interview of local guides, local people, and the staff of travel agencies. Collection of maps on tourist areas and guidebooks for Chinese domestic tourists.
	March 16–19, 2017	Observation of exhibitions on earthquake ruins and museums in Beichuan County and New Beichuan. Interview of local guides, local people, and the staff of travel agencies. Collection of Chinese newspapers and magazines with articles on the Sichuan Earthquake.
	March 9–11, 2018	Observation of exhibitions on earthquake ruins and museums in Beichuan County and New Beichuan. Interview of a manager and staff of the museum, a government official of New Beichuan, local guides, local people, and professionals from Southwest Jiaotong University and Tongji University. Collection of statistics on the number of visitors arriving at each facility.
Kathmandu (including Patan), Nepal	November 15–19, 2017	Visit of two traditional houses in Kathmandu (Patan and Kirtipul). Interview of three owners of traditional houses in Kathmandu (Patan, Kirtipur, and Bhaktapur). Participation in the Patan Workshop, facilitated by Prof. Takeyuki Okubo (Ritsumeikan University), and interview of the workshop's facilitators.
Lijiang, China	August 10–14, 2016	Fieldwork in the central area of Dayan Old Town, including the measurement of the street widths using 114 points on 34 streets and the carrying out of a survey of 12 characteristic open spaces. Interview of the dean of the Mu Fu Museum and the staff of the World Heritage Lijiang Old Town Protection and Management Bureau. Analyses of the data collected from satellite images.
Bergama, Turkey	September 2–5, 2015	Fieldwork in water-supply sites and on daily activities in the historic town. Interview of the mayor, the staff of the UNESCO department of the Bergama city government, and local people. Collection of management data on *çeşmes* (water-supply facilities), street map, tourism map, and new transportation-route map for spatial analysis.
	February 15–17, 2018	Fieldwork in four disaster-mitigation squares in Bergama. Interview of a fireman at the Bergama fire station, the mayor, the staff of the UNESCO department of the Bergama city government, and local people.
	February 14–15, 2019	Organization of an international professional meeting and a community-based workshop, along with Ms. Yasagül Ekinci (who worked at the UNESCO department of the Bergama city government at the time).

extensive damage. Its seismic center was in Ying Xiu Town, Aba Zangzu Qiangzu, which is an autonomous prefecture located 70 km from Chengdu, the capital of the Sichuan Province. In the following subsections, we examine the reconstruction process following this disaster.

3.1.1 Quick recovery

In September 2008, the Chinese government issued "the state overall planning for Post-Wenchuan Earthquake restoration and reconstruction" (General Office of the State Council of the People's Republic of China, 2008). Only two and a half years after the earthquake, the government declared the achievement of the reconstruction. Here, the term "achievement of the reconstruction" implies that the livelihoods and economic status of the survivors were restored to a higher level than the one existing before the disaster. The government disclosed that the counterpart-assistance method facilitated a quick recovery because each province and directly controlled municipality had competitively offered reconstruction support. However, the excessive development in the region resulted in the rapid construction of new towns, which forced earthquake survivors to move to different places.

3.1.2 Large-scale emigration and the construction of New Beichuan

From September 2009 to February 2011, large-scale emigration occurred in the capital of Beichuan Qiang Autonomous County. As Beichuan County is located on an active fault, it suffered serious damage during the earthquake. On the tenth day following the earthquake, the then Chinese Premier Wen Jiabao visited the location and opined that it was worthwhile to preserve the site as an earthquake ruin and transform it into a related museum. Subsequently, the government decided to maintain this area in its existing state and transfer the county's capital elsewhere (Fig. 2.2A). Accordingly, "New Beichuan" was constructed in a part of the county that was 20 km away from the original capital and that had a gross area of 7 km^2. Beichuan County is the only autonomous county inhabited by the Qiang ethnic group in China. Therefore, New Beichuan was built as a modern city in the Qiang architectural style. Furthermore, in this area, a touristic landscape was constructed to promote postdisaster industrial development (Fig. 2.2B). Over time, New Beichuan has evolved into a region that reflects the "success" of postdisaster relief and reconstruction efforts and the "strong leadership" associated with public policies in China.

Figure 2.2 Earthquake ruin in Old Beichuan County Town (A) and the touristic landscape of New Beichuan (B). Takae Tanaka.

3.1.3 National resilience and social places

The national authorities quickly reconstructed New Beichuan, and hence, the new residents of this area could not involve themselves in the recovery process. Consequently, various problems occurred in their daily lives. First, New Beichuan had two different types of residents: locals who had been living in the area before the earthquake, and people who had lived in "Old Beichuan" and moved to the new town. These two ethnic groups were divided among four residential districts, and previous research clarifies the conflicts that existed between the two groups (Qiu, 2017). After the earthquake, the local community, comprising Han Chinese farmers, was deprived of their homeland and witnessed their land being reconstructed to promote the Qiang culture. On the other hand, the government compensated the local people for their losses by providing them with new houses for free. However, the people from Old Beichuan who lost their homes had to purchase the new houses by themselves. Second, the random housing allocation caused the disintegration of preexisting social relationships. As a result, relatives and friends had to live separately and were distributed among 154 buildings and 3638 rooms (Beichuan xian Renmin Zhengfu, 2010). Moreover, some people were not satisfied with their rooms. Third, the population was smaller than the number anticipated in the government plan, which considered 68,000 residents (Chen, Wei, Zhu, & Sun, 2011).

Therefore, although the government affirmed its national resilience to disasters through the quick reconstruction of New Beichuan, the survivors faced several social and cultural problems when adapting to the changed situation. Even today, the new residents are adapting to their social places and transforming this strange land into their home in the middle of ongoing recovery efforts.

3.2 The reconstruction process following the Gorkha Earthquake

In this subsection, we turn to the disaster experiences and reconstruction process associated with the Gorkha Earthquake in Nepal, which occurred on April 25, 2015. This event had a magnitude of 7.9 on the Richter scale, and its epicenter was approximately 105 km away from the Kathmandu Valley (central Nepal). According to reports, the relief operations implemented after the disaster were quite encouraging. However, rehabilitation and reconstruction were delayed due to political, institutional, legal, and governance issues. For instance, the reconstruction process increased the number of people living in poverty from 700,000 to 982,000 (UNICEF, 2015).

Patan, one of the World Cultural Heritage cities located in the Kathmandu Valley, has many ancient temples, traditional houses, and royal palaces, and these were still being reconstructed in November 2017—that is, two and a half years after the earthquake. In the old cities of Patan, Bhaktapur, and Kirtipur, the community courtyard gardens called *chowks* are densely surrounded by middle-sized apartment dwellings (having about four to five floors), in a way that the entrances of such dwellings face the *chowk*. Moreover, narrow alleys connect the *chowks* under the dwellings (i.e., on the first floor) (Hino, Okazaki, & Ochiai, 2014). Hence, several public spaces are continuous and intricately connected in the old district. In this manner, social places such as ancient temples, narrow alleys, and *chowks* overlap each other and are frequently used by the residents in their daily lives. In addition, despite being still under restoration, these social places are actively used by visitors and tourists.

3.2.1 Quality of the houses

There is a direct association between the quality of the outer walls of the dwellings and households' economic standing. Affluent households are more likely to wall their housing units with cement, concrete, or cement-bonded bricks or stones, whereas poor households tend to wall their units with mud-bonded bricks, stones, wood, or branches (Central Bureau of Statistics, 2011). According to an interview with Suraj Pradhan, an architect/researcher who analyzed the housing situation after the Gorkha Earthquake in a historical site, the materials used in the dwellings' outer walls strongly affected the damage levels of the houses. In most cases, the walls of the damaged dwellings were made of mud-bonded bricks and stones.

Further, interviews with house owners in Patan, Kirtipur, and Bhaktapur revealed that although there was a government support system for the damaged houses, people did not use it. They argued that contacting this support system was a time-consuming activity and required the completion of many application forms. They also claimed that even after contacting the government, there was no guarantee that the residents would get adequate support. Hence, people attempted to build their houses by themselves with as little money as possible, without the technical assistance of architects. Furthermore, the adopted restoration method aimed at covering buildings' cracks with a simple cement coating. Hence, it did not foster resolving the structural problems of the dwellings. We identified these types of restoration by the owners of the damaged house in the fieldwork conducted in Patan and Kirtipur. Such restoration merely hides the problem and does nothing to improve the resistance of such buildings to future earthquakes.

3.2.2 Patan workshop to update the community-based disaster risk management plan

One of the authors of this chapter (Kano) joined the Patan workshop facilitated by Professor Okubo from Ritsumeikan University in 2017 (Ritsumeikan University, 2016). The workshop project actually started in 2008—that is, before the Gorkha Earthquake—for the development of the disaster risk management (DRM) plan in the old residential district of Patan, which is located on a buffer zone in the World Heritage site. In the 2017 workshop, representatives of the local community updated the community DRM map to mark evacuation routes from their houses through *chowks*. Simultaneously, they elaborated the DRM Activity Guideline (Takasugi, 2017).

3.2.3 Social places and activities in times of disasters

From the interviews conducted as part of the fieldwork in November 2017, we identified the following aspects regarding local disaster experiences. First, the semi-open space called *pati*, which is used daily by the residents, is very useful in times of disasters. Although the *pati* was originally constructed for the convenience of tourists, it was recently used by pilgrims and local people as public meeting places (Fujioka, 1992). In

Figure 2.3 The social place (*pati*) used for laundry services (A) and the social activities taking place in a *chowk* (B). Tomoko Kano.

Bhaktapur, a World Heritage site that was severely damaged by the earthquake, residents used *patis* and its surrounding places for laundry services (Fig. 2.3A). These social places can adapt themselves to emergency situations for the benefit of both locals and tourists.

Second, some of the residents' daily activities are beneficial during emergencies. For instance, the community gardens called *chowks*, located inside historical houses, functioned as places for cooking and food supply for the local people after the earthquake. Fig. 2.3B depicts the social activities in a *chowk* that was used as a cooking place for Patan workshop attendees. The workshop and other disaster-preparedness initiatives gave the residents a critical opportunity to "build back better" by integrating disaster risk reduction into community efforts and, thereby, making the community more resilient to future disasters. In addition, such efforts gradually enhance residents' disaster awareness, which, in turn, helps to keep the town safe.

3.3 The disaster risk reduction potential of social places in the Old Town of Lijiang

In this subsection, we examine the potential of social places and activities in disaster risk reduction in the Old Town of Lijiang, which is located in southwest Yunnan and was designated as a World Heritage site in 1997. The town has three sections, including Dayan Old Town. The World Heritage property area in Dayan Old Town spans 110 ha, including Heilongtan Pool. Once an important commercial and strategic town, the Old Town of Lijiang retains its historic townscape even today.

3.3.1 Spatial characteristics

Dayan Old Town has many waterways and streets. Three waterways run through the town and its main streets, and their branch alleys radiate from the central square. Along these waterways and streets on undulating land, two-story houses, generally with wooden frames, brick or soil walls, and roof tiles, have been constructed. These houses enclose individual courtyards. Therefore, in this town, waterways and the houses along narrow alleys, or curved or pitched streets, form a unique townscape.

A spatial analysis of the central area of Dayan Old Town reveals the following points (Nakamura & Gota, 2017). First, according to network analysis, some streets near the central square are highly valued in the centrality index, and people are potentially concentrated in these areas. Second, we identified, based on the 114 measure points selected from 34 streets at random, that most of the streets are very narrow (with width varying between 2 and 4 m). Third, the measurement of 12 characteristic places on the streets reveals streets and/or waterways surrounded by houses, a small plaza, and small irregularly shaped open spaces, including famous public wells called "one pool of water" and "three pools of water" (Kuroyanagi, Ichikawa, Xuguan, & Suzuki, 2012). Small open spaces exist along the narrow streets in the old town (Fig. 2.4).

3.3.2 Transformation of the town

Since its designation as a World Heritage site, the Old Town of Lijiang has undergone significant changes resulting from the rapid tourism development of the area (Tourism Development Committee in Lijiang, n.d.). However, the region's spatial characteristics were maintained while facilitating its transformation from a residential area into a tourist site. As a result of such changes, the Naxi ethnic group, who had lived there previously, was forced to move out of the town, and non-Naxi outsiders have moved in to work in the tourism industry. A Naxi woman tour guide stated that she went to work in the old town every day and that the Naxi ethnic group did not live there anymore. The town's Naxi population decreased over the years, and today the community has almost disappeared. Yet, non-Naxi outsiders who have moved in are probably less aware of disaster risk prevention than the Naxi people. In the absence of its original residents, it is difficult to prepare the town's community for disaster risk reduction.

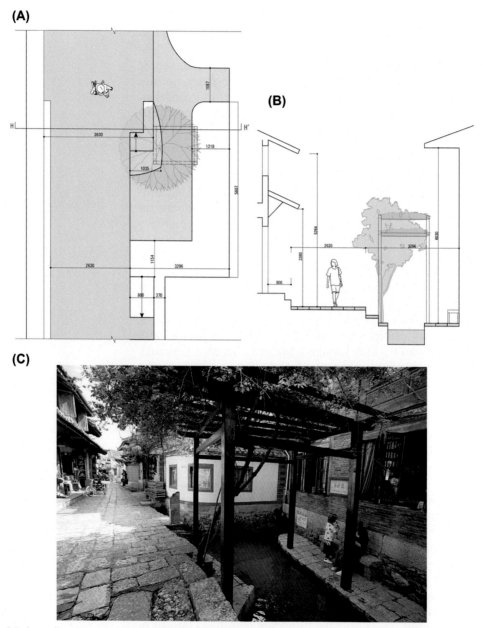

Figure 2.4 A small open space by the side of a waterway: (A) plan, (B) section, and (C) photograph. Momoyo Gota.

3.3.3 Risk reduction potential of social places and social activities

The Old Town of Lijiang is earthquake prone, and a major event occurred in 1996. It damaged local buildings, including those in Dayan Old Town. Furthermore, Dayan Old Town faces the risk of urban fires because its houses are made of wooden frames. Indeed, several houses were destroyed by fire in 2011 and 2013.

Dayan Old Town has numerous narrow roads; further, it has small open spaces that are physically effective in reducing fire damage. Moreover, residents use spring water in public wells. Although the town's population has declined over the years, people, including commercial workers, still use these places, which are enjoyed by tourists as well. These social places, which conserve water and connect residents, commercial workers, and tourists, contribute to disaster risk reduction.

In addition, local people have long had the habit of damming the river and cleaning the central square using the flooded water. During the fieldwork, we observed how a street was hosed down with water (Fig. 2.5). Although the original resident community has disappeared from the town, the use of fire hydrants and hoses as part of the daily routine is expected to improve the nonresident commercial workers' awareness of disaster risk reduction.

Figure 2.5 Washing a street with water in Dayan Old Town. Momoyo Gota.

3.4 Proposal of a disaster risk mitigation plan for Bergama

In this subsection, we propose a disaster risk mitigation plan for residents and tourists in Bergama (Pergamon), Turkey, as a model of a living-heritage town. Although Bergama is located in a seismic zone, it has not experienced any major earthquakes in recent years. However, disaster preparedness enables the integration of disaster risk reduction into development measures and the analysis of the potential of social places in heritage towns. We formulated the Bergama plan in due cooperation with the local government and following the identification of major issues related to the existing disaster mitigation plan.

3.4.1 Bergama as a living-museum city

Bergama is an Anatolian city located in Izmir Province. This living-museum city is an important cultural and tourism center in the Aegean area that includes several valuable archaeological and historical sites. It has survived the periods of Hellenism, the Roman Empire, the Byzantine Empire, the Ottoman Empire, and the Turkish Republic (Pirson, 2015). In 2014, Bergama was listed as a World Cultural Heritage site—"Pergamon and its multilayered cultural landscape."

Today, the main sites of the ancient Kingdom of Pergamon are located north and west of the modern city of Bergama. The notable ruins in Bergama are the Acropolis (a fortified hill), the sacred Asklepion (an ancient medical and health complex), and a Roman Temple called Kızıl Avlu (meaning Red Basilica or Red Courtyard), which was converted into a church in the Byzantine period, then the north of the main structure was turned into a mosque by the Turks, being this its current use (Sahin, 2000). In addition, the Ottoman Period witnessed the creation of a renewed water system and several commercial and public buildings, such as the Turkish public baths (*hamams*), water-supply facilities (*çesmes*), and bazaars (*arastas*), which enhanced the city's social life. Some of these buildings and facilities have been in use up to today.

3.4.2 Spatial characteristics and attractive features of social places

The site of the disaster risk mitigation plan is the residential district (approximately 30 ha) in the core zone (332.5 ha) of the World Heritage site, on the side of Kale (castle) Hill, which is 330 m above sea level. The traditional houses in this district have generally two stories and are made of stone, brick, and timber frames. Today, among the traditional houses in Bergama, only 12% are in good condition and 48% are in average condition. Conversely, 22% of the houses are in bad condition and in need of repair and rehabilitation, which is more than the preliminary estimates (Bergama Belediyesi, 2008). An interview with a local fireman revealed that, nowadays, fires occur approximately 200 to 250 times a year in Bergama and its surrounding villages. This is due to house roofs being largely made of timber.

The residential district includes narrow streets and dead-end alleys. The alleys form mazes, as most of them are not connected perpendicularly to the streets on the hillside. Therefore, tourists and visitors sometimes get lost in the maze, although tourists' maps provide adequate information. However, the occurrence of the spontaneous spaces in the alleys offers a coziness experience to both residents and tourists. Jinnai and Arai (2002) discussed the space of the residential district as follows: The inhabitants who have lived for a long time in the labyrinth-like space can clearly detect its differences. They are familiar with the entire maze-formed space, rather than just the simple planning space. On the other hand, disasters such as the 1842 flood and the 1853 fire damaged the traditional urban fabric. In the fire, nearly 400 shops, 200 houses, and 5 bazaars were burnt down (Binan Ulusoy & Binan, 2005). Hence, it is desirable to maintain the attractive features of the maze-formed space while planning for disaster risk management.

3.4.3 Social places and activities of the living-heritage town

Here, we present an example of social places of this living-heritage town and some social activities by addressing the following questions: Which places are cultivated as social places in Bergama? How can social places be used in emergency situations?

The first answer concerns the water-supply facilities. The local people in Turkey use public-water facilities and perform daily routines as follows: mosques, for example, have a water-supply system that provides water for religious purification. Meanwhile, some *çesmes* located in the residential district are used in everyday life. *Çesmes* were sometimes built by affluent residents in front of their houses, and their neighbors often gathered around these areas. Hence, these places can be considered to function as a type of community base. Currently, the UNESCO department of the Bergama city government records these *çesmes* as cultural assets. However, the idea of utilizing *çesmes* as tourism resources has not yet been considered, although they are part of

Bergama's living heritage. This is because *çesmes* are a common feature of Bergama's townscape. Furthermore, if a fire occurs in a densely populated residential district structured like a maze, the water from the *çesme* can be used for the initial extinguishment operations. The *çesmes* themselves can also be used as evacuation lifelines in the event of an earthquake.

The second answer concerns the daily activities of men and women. In Bergama, men and women differ in the manner in which they take rest and spend the day. Fig. 2.6A depicts the spontaneous spaces that men use for chatting—that is, drinking under trees in front of shops. Today, such social places are rapidly being destroyed in the name of urban development. In their turn, women often stay in the residential districts, such as the maze, and interact with their children and neighbors in front of the gates of traditional houses (Fig. 2.6B). These social places and residents' activities keep the town safe by enabling them to share information and sustain their communities.

Finally, the third answer highlights the different uses of social places. In the Ottoman Empire, the Turks, the Greeks of Anatolia, the Armenians, and the Jewish people lived in the same district. In 1899, Bergama's total population was 23,590 (Örnek Özden, Yerliyurt, & Seçkin, 2006). Among these, 17,139 were Turks, 3585 were Greeks having Turkish citizenship, 281 were Armenians, 495 were Jewish, and 74 were from other minority groups. Today, people of different ethnicities still live in the same area. In addition, Bergama is an attractive tourism destination, receiving people from European, Asian, and other origins who use the same public space. Yet, the tourists have different styles of using social places. Therefore, we should understand the different uses of social places, particularly concerning emergency situations.

Figure 2.6 Typical daytime social activities of men (A) and women (B) in Bergama. Tomoko Kano.

3.4.4 An upgraded disaster risk mitigation plan for Bergama

Currently, a disaster risk mitigation plan developed by AFAD-Izmir (the disaster- and emergency-management authority in Izmir) is operational in Bergama. However, the citizens of Bergama are not aware of this plan. Taking this into account, we propose the integration of social places into the disaster risk mitigation plan.

As the evacuation square has been currently designated for the district alone, we first had to select the evacuation route. We proposed the creation of an evacuation route that extended from the residents' houses to the evacuation square and connected the social places in the residential district (Fig. 2.7). The social places of the heritage town that are used daily by locals—namely the *çesmes*, mosques, boutique hotels, and *hamam*s—function as disaster lifeline facilities and information centers. Therefore, we integrated these social places of the living-heritage town in the evacuation route. Furthermore, we suggested separating places for women and men in the evacuation square because they engage in different social activities. Furthermore, nursery spaces and babies' rooms should be placed next to women's places. To increase the local people's awareness of disaster risk preparedness, we conducted a community-based workshop in Bergama as part of our research.

3.5 Outcomes of the professional meeting and the community-based workshop in Bergama

The last step in this research project was the organization of a community-based workshop in Bergama in February 2019, in collaboration with Ms. Ekinci and her colleagues, to enhance the local community's disaster risk awareness. This activity was conducted as a follow-up to an earlier workshop held in Patan, facilitated by Professor Okubo.

3.5.1 Professional meeting

On February 14, 2019, we organized an international professional meeting in Bergama's city-hall building. In this meeting, a representative of the fire station, a professor from Ege University (Department of Tourism), a researcher from AFAD-Izmir, a representative of Bergama Museum, and the city's mayor participated and discussed several issues. The presentation revealed that the clarification of various kinds of disaster risks in the living-heritage town should be prioritized. Indeed, besides the risk of fire and earthquake, the flooding of the Selinos River—

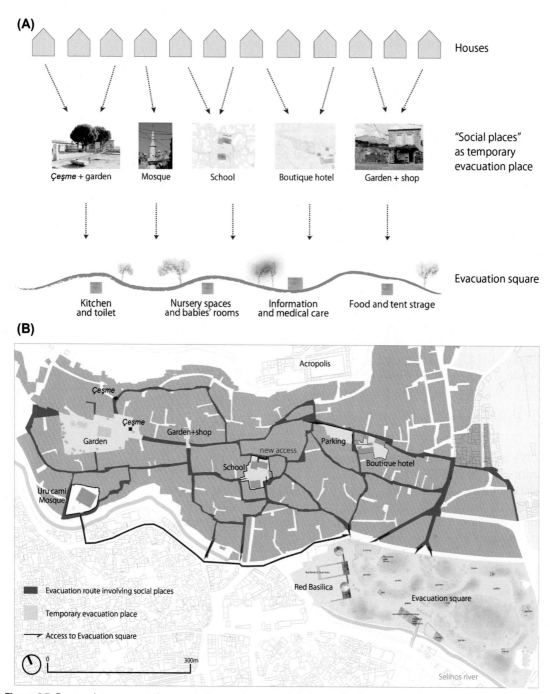

Figure 2.7 Proposal to connect the evacuation routes to social places (A: diagram; B: plan). Tomoko Kano and Taeko Nakatsubo.

which flows through the center of a historical area—and the possibility of landslide occurrence—historical dwellings located on the hillside have the possibility of being damaged by rocks falling from Kale Hill—were matters of concern. Then, the participants shared the situation of their professional research activities and got their hints for solutions to their individual problems.

In the meeting, we presented our research projects on the disaster experience and recovery scenario in China, the disaster experience and recovery process in Nepal, and then we shared some details of our local activities. In addition, we proposed the development of a disaster risk mitigation plan for Bergama, from an outsider's perspective.

Throughout the meeting, we emphasized the importance of formulating disaster risk management plans, particularly in a World Cultural Heritage site such as Bergama. According to a report from Ataberk (2014), approximately 350,000—400,000 people visit the museums and ruins in Bergama each year. Therefore, such a tourist city must prepare emergency food and water supplies, transportation-system capacity, medicine, tents, and emergency and information-sharing facilities for more than the number of the local population. The disaster risk management plan for a tourist city area must be distinct from the one of its surrounding countryside that does not involve intense flows of people. To create such plans for a tourist area, one should consider the area's population, conservation regulations, and spatial development.

Furthermore, we emphasized the necessity of rescue teams having professional knowledge on possible damage to the cultural assets. The rescue team should have relevant knowledge and be able to apply techniques to protect both human life and heritage assets while saving survivors. Obviously, lifesaving activities should be prioritized over other ones. However, heritage assets are often needlessly damaged when survivors are saved due to the rescue party's lack of knowledge about them. We also observed that residents' everyday lives will continue to depend on the region's heritage assets even after a disaster, and hence such assets should be also safeguarded as much as possible by the rescue teams.

3.5.2 The community-based workshop

On February 15, 2019, we organized the community-based workshop in Bergama's city hall. The workshop was attended by nine people, including men and women who were engaged in local tourism, educational, and cultural activities and had

Table 2.2 Attendees of the community-based workshop organized in Bergama.

Attendee number	Position and affiliation	Gender	Workplace
1	President of the Bergama Women's Cooperative	Female	Historical site
2	Owner of a tourism agency in Bergama	Male	Historical site
3	Owner of one of the first boutique restaurants in Bergama (most of the meals are local foods)	Male	Close to the city center
4	Teacher and director of a primary school	Male	On the hillside of the Acropolis
5	Owner of an olive oil shop and a researcher of Bergama's local history	Male	Near the *arasta*, historical site
6	Civil engineer who provides guiding services in Bergama	Male	City center, historical site
7	Restoration worker and owner of a handmade gift shop in Bergama	Male	*Arasta*, historical site
8	Owner of a cafeteria	Female	*Arasta*, historical site
9	Owner of a pharmacy	Female	Outside the historical site of Bergama
10	Architect at the UNESCO Department of the Bergama city government	Male	City center, historical site
11	Archaeologist (MA) at the UNESCO Department of the Bergama city government	Male	City center, historical site

maintained each activity as community leaders in their daily lives (Table 2.2). In summary, they were people of influence in the neighborhood, although the chairman of the neighborhood association, called *mukhtar*, could not participate in the workshop.

The workshop attendees were separated into two groups, and each group included one facilitator who had worked at the UNESCO Department of the Bergama city government. We presented to the groups the same disaster risk management topics, particularly related to earthquakes and fires, and they discussed each of these from the perspective of a community leader and that of a local citizen. These community leaders proactively requested their effective involvement in the formulation of a disaster mitigation plan.

3.5.3 Major topics discussed at the community-based workshop

In this workshop, the participants discussed risk management systems for disasters such as earthquakes and fires, the characteristics of different kinds of risk, daily activities in the eventuality of a disaster, actions to be performed after the disaster, and life at refuges. In addition, they mapped out a dangerous area in the Bergama living-heritage town while examining the evacuation route at the time of a disaster. This process helped to improve the participants' disaster-prevention awareness. Furthermore, they wanted to provide information on the evacuation square as soon as possible because a place of refuge had not yet been identified, and they demanded that the Bergama city government starts disaster-mitigation activities at the earliest.

One of the impressive discussions held during the workshop was on the lifesaving actions and the management of cultural heritage during disasters. Initially, the community leaders remarked that the topic should be discussed at the national level since they could not access cultural heritage without the national government's permission. Hence, they considered this topic to be *"somebody else's problem."* However, a staff member of the UNESCO Department of the Bergama city government (attendee number 11) pointed out that the protection of a World Heritage site in Bergama is directly related to the residents' lives, particularly those working in the tourism industry. This statement abruptly revised the community leaders' thoughts on the matter. The discussion organically developed by addressing the following questions: How do we prevent the theft of cultural heritage at the time of a disaster? What types of systems and education are required in community organizations to prevent the theft of cultural assets?

Another important topic discussed during the workshop was the following: If part of the cultural assets unfortunately collapses during an earthquake while people are being rescued, how can one reduce the damage to the remains? The community leaders were surprised to learn that, once they collapse, cultural remains cannot be completely restored. Subsequently, they commented after the discussion that causing damage to the World Heritage site and leaving this damage to be solved by future generations is not a mission of Bergama's residents.

3.5.4 Results of the workshops

Ms. Ekinci commented that

the professional meeting and community-based workshop were successful and fruitful. These are the activities that I have wanted to conduct in the World Heritage site of Bergama for many years. I was impressed by the community leaders, who assumed their responsibility with sincerity, despite the presence of difficult discussion topics. On the other hand, we, staff at the UNESCO Department of the Bergama city government, were surprised by some of the topics suggested by [the workshop organizers], as follows: Should men and women share the same evacuation site and space? Is it acceptable if there is the same evacuation space for residents and tourists? These topics seemed to be unexpected not only for community leaders but also for us, and we felt that these topics broadened the scope of our discussions.

The eye-opening meeting concluded that in the case of World Heritage sites, a shared commitment to a disaster risk mitigation plan can save human lives, heritage, and livelihoods in a joint effort. Currently, a disaster risk management committee is expected to be established in the local community, with the support of the Bergama city government. If realized, this may be the first of such efforts in Turkey. Furthermore, the contents of the international professional meeting and community-based workshop will possibly be reported to the UNESCO headquarters by the UNESCO department of the Bergama city government. These disaster-management activities have strengthened the local community's resilience to disasters.

4. Enhancing community resilience and promoting sustainable heritage tourism

Our study showed that there was a sharp contrast between the recovery from the Wenchuan and Gorkha earthquakes. This concerned the regions' social places and residents' social activities, such as the implementation of a large-scale postdisaster reconstruction process versus a community-scale recovery plan, the emphasis on replacement versus rebuilding, and the realization of quick versus slow recovery. By examining these disaster experiences, this chapter revealed the potential of social places and activities in disaster risk reduction in the Old Town of Lijiang. It pointed out that social places, such as small open spaces and pools, and the related daily routine activities of local people, have meaningful roles in times of disasters. Particularly, in World Heritage towns, it is not easy to build new evacuation buildings or widen streets, due to the presence of strict regulations for traditional buildings. Hence, traditional social places, such as water-supply facilities and their surrounding areas, as well as residents' social activities, especially daily routines, are important for disaster risk mitigation. This argument is possibly valid not only in the World Heritage sites of Lijiang, Patan, and Bergama but also in other heritage towns.

These social places and activities are sometimes suitable for the application of measures related to disaster risk management, as well as for building community resilience. Therefore, residents, commercial workers, and tourists should be integrated into and allowed to contribute to the towns' efforts toward disaster risk reduction. Furthermore, we proposed the integration of social places and activities of the living-heritage town in the disaster risk mitigation plan for Bergama. Also, we should evaluate the effectiveness of the community-based workshop of Bergama, and then share and design the method of disaster education required by community organizations.

We sustain that the joint efforts by community leaders and residents will contribute to the enhancement of community resilience and the promotion of sustainable heritage tourism. This can enable the development of disaster risk mitigation plans through collaborations between the government and local communities, as called for by the third priority of the Sendai Framework.

Acknowledgments

This chapter is based on the research project *World Heritage and Disaster Risk Mitigation: For Sustainable Heritage Tourism in Asia,* carried out from April 2016 to March 2019 by T. Kano, S. Yamashita, M. Gota, M. Doshita, T. Tanaka, and H. Iwahara, and supported by the Japan Society for the Promotion of Science KAKENHI—grant number 16H03332. We acknowledge the works of Ms. Yasagül Ekinci and Mr. Fatih Kurunaz in organizing the professional meeting and workshop, as well as providing valuable comments and first-hand data. We extend our special appreciation to Ms. Naoe Shiigi Sen, for supporting and guiding us in the fieldwork and the workshop conducted in Bergama, and to Ms. Taeko Nakatsubo, representative of the Japanese architectural design office ON, for cooperating with the design of the disaster risk plan of Bergama.

References

Ataberk, E. (2014). Assessment of tourism effects on geographical area from the perspective of local tourism actors: Bergama case (Izmir/Turkey). *European Journal of Geography, 5*(2), 27—42.

Beichuan xian Renmin Zhengfu. (2010). *Beichuan Nianjian [Beichuan Yearbook]* (in Chinese).

Bergama Belediyesi. (2008). *Bergama Koruma Amaçli Nazim Imar Plani [Bergama city conservation plan].* Ankara: Egeplan Planlama (in Turkish).

Binan Ulusoy, D., & Binan, C. (2005). An approach for defining, assessment and documentation of cultural heritage on multi-layered cities, case of Bergama

(Pergamon)—Turkey. *Proceedings of the Scientific Symposium, ICOMOS 15th General Assembly and Symposium, Xi'an, China.* Retrieved from http://openarchive.icomos.org/id/eprint/275/1/1-8.pdf.

Central Bureau of Statistics. (2011). *Nepal Living Standards Survey 2010/11 Statistical report volume one.* Retrieved from https://nepalindata.com/resource/nepal-living-standards-survey-2010-11statistical-report–volume-one.

Chen, Z., Wei, W., Zhu, Z., & Sun, T. (2011). A case study on sustainable planning practice in Beichuan New Town. *City Planning Review, 35,* 31–36 (in Chinese).

Fujioka, M. (1992). *Nepal kenchiku shō yō [Essays on Nepalese architecture].* Tokyo: Shokokusha Publishing (in Japanese).

General Office of the State Council of the People's Republic of China. (2008). *Wenchuan dizhen zaihou huifu chongjian zongti guihua [The state overall planning for post-Wenchuan Earthquake restoration and reconstruction]* (in Chinese).

Hino, Y., Okazaki, K., & Ochiai, C. (2014). Study on living styles of current dwellings and disaster awareness of people in Patan, Kathmandu, Nepal. *Reports of the City Planning Institute of Japan, 12,* 160–163 (in Japanese).

Jinnai, H., & Arai, Y. (2002). *Isram Sekai no Toshi Kuukan [Urban space of the Islamic world].* Tokyo: Hosei University Press (in Japanese).

Kuroyanagi, A., Ichikawa, T., Xuguan, S., & Suzuki, N. (2012). A research study on residents and water use in Lijiang old city, Yunnan province, China: Transformation in using the water of "three pools of water." Part 1. *Journal of Architecture and Planning, Architectural Institute of Japan, 77*(672), 359–367 (in Japanese).

Nakamura, M., & Gota, M. (2017). Spatial characteristics of the old town of Lijiang focusing on street networks. *Summaries of Technical Papers of Annual Meeting, Architectural Institute of Japan,* 1061–1062 (in Japanese).

Örnek Özden, E., Yerliyurt, B., & Seçkin, E. (2006). Continuity of the spirit of a place case of a historic town. *42nd ISoCaRP Congress, Istanbul, Turkey.* Retrieved from http://www.isocarp.net/Data/case_studies/800.pdf.

Pirson, F. (2015). *Pergamon: A Hellenistic capital in Anatolia.* Istanbul: Yapi Kredi Yayinlari.

Qiu, Y. (2017). Mo sheng de Xin Jia yuan: Yi di Zhong jian hou Xin Beichuan Jumin de Kongjian Shangque he Wenhua Tiaoshi [A strange new home: Negotiation and cultural adaptation of New Beichuan residents after remote reconstruction]. *Journal of Southwest University for Nationalities (Humanities and Social Science), 3,* 32–39 (in Chinese).

Ritsumeikan University. (2016). *Radiant Issue #2: Living with a disaster.* Retrieved from http://www.ritsumei.ac.jp/research/radiant/eng/disaster/story1-1.html.

Sahin, B. (2000). *Pergamon in the history of Anatolia.* Bergama: Çağdas Matbaacılık.

Takasugi, S. (2017). *Disaster risk management plan with assessment of traditional resources for disaster mitigation and response.* Retrieved from http://r-dmuch.jp/jp/researchnews/contents/17.12.02.Disaster_Mitigation_Design_for_Historic_Cultural_Cities.pdf.

Tourism Development Committee in Lijiang. (n.d.). *Official website of Lijiang Municipal Administration of Culture and Tourism.* Retrieved from http://www.ljta.gov.cn/html/infor/tongjixinxi/lytj.html.

UNICEF. (2015). *Nepal earthquake emergency fundraising 26th report.* Retrieved from https://www.unicef.or.jp/news/2015/0228.html (in Japanese).

3

Investing in contingency in a heritage site

Liliane Hobeica[1] and Adib Hobeica[2]

[1]RISKam (Research Group on Environmental Hazard and Risk Assessment and Management), Centre for Geographical Studies, University of Lisbon, Lisbon, Portugal; [2]Independent Consultant, Coimbra, Portugal

1. Built heritage at risk

The Sendai Framework for Disaster Risk Reduction 2015—2030 calls for investing in disaster risk reduction (DRR) as a key priority, although not prescribing the actual means, types, and timings of such long-term resource-commitment decisions (UNISDR, 2015). These issues are particularly important when it comes to heritage assets—outstanding and valuable cultural relics that have been resisting time. When located in hazard-prone areas, their heritage status relates not only to their intrinsic historic values and meanings but also to their role as reminders of living with the latent risks. This chapter explores the integration of DRR into the management of heritage sites, aiming to highlight the related intricacies. In particular, we examine the crucial interplay of physical and sociocultural dimensions, which might be overlooked by the Sendai Framework.

Our study focuses on the relationship between floods and the Monastery of Santa Clara-a-Velha, originally built in the 14th-century on a small plateau near the Mondego River's left-bank shoreline in Coimbra (Portugal). This complex relationship took varied forms across the centuries, as illustrated by the successive resort to resistance, resilience, retreat, and resignation strategies. Having partially withstood several flood events during its lifetime, this historic monument was the target of a requalification project at the beginning of the 2000s, following the conclusion of the Mondego regulation works in the 1980s. Indeed, this flood-mitigation intervention at the basin scale prompted the city of

Investing in Disaster Risk Reduction for Resilience. https://doi.org/10.1016/B978-0-12-818639-8.00005-3

Coimbra to rediscover its riverbanks as both a primary location for leisure and recreation activities and a contemporary landmark. The cultural entities in charge of the Monastery of Santa Clara-a-Velha also took this opportunity to recover the remains of this remarkable religious and royal complex.

Our research intended to highlight the prevalent flood-risk culture underlying the most recent intervention (and related investment decisions) in this monument site. We also aimed to illustrate how architectural design can function as a DRR tool to support investments toward the resilience of both a built heritage site and the concerned community. To this end, we adopted the case-study method (Yin, 2009), as it is geared toward reaching a broad view of a given phenomenon—in this case, flood adaptation through design in a heritage site. This method is particularly useful in the architectural domain, for it supports scrutinizing concurrently the design's context, product, and process, thus providing a holistic apprehension of the built output (Foqué, 2010). In terms of procedural steps, we collected data through desk review, in situ observations, and semistructured interviews with key stakeholders: three architects, three engineers, two landscape architects, and one archaeologist (the interview quotes appear in italics throughout the text). Our analyses comprised the elaboration of a "thick description" of the case (Dawson, 2010), taking flood adaptation as our research lens.

In Section 2, we discuss some key concepts and perspectives related to risks and the built heritage. We then introduce, in Section 3, the geographical and sociocultural contexts in which the examined requalification intervention evolved. In Section 4, we present the intervention process, comprising the related decisions in terms of both the management of flood risk and the design stance vis-à-vis the monument. This is followed in Section 5 by a description of the resulting design output and our interpretations of its sociocultural repercussions. In that section, we also highlight the consolidation of the prevalent flood-risk culture, which could itself have been also *requalified* along with the built heritage. In Section 6, we delineate some final remarks about the integration of DRR into the management of heritage sites and the overall mindset of the Sendai Framework's third priority.

2. Fragility and resilience of the built heritage

According to Choay (1992), the Western notion of historic monument is associated with the need of keeping alive the elements of the past that support the cultural identity of a given society. Indeed, the heritage designation of built structures with outstanding material and immaterial values has frequently been induced by the imminent loss of these assets and the related repercussions in sociocultural terms. Whereas their perceived transience is a key characteristic of these cultural assets, their preservation is a potentially more pressing issue when they are located in hazard-prone settings given their increased risk of disappearance.

On the other hand, conservation and restoration theorists argue that in many cases the built heritage supports resilient practices of the communities in which it is located (Jigyasu, 2016). Accordingly, international organizations often take the protection of heritage as a promoter of societies' resilience (ICOMOS-ICORP, 2013). The International Centre for the Study of the Preservation and Restoration of Cultural Property, the International Council on Monuments and Sites, and UNESCO joined efforts with the United Nations Office for Disaster Risk Reduction to produce a dedicated study in this regard (UNISDR, 2013). The recommendations ensuing from this strategic document reverberated in the Sendai Framework, whose first and third priorities for action include among their concerns the evaluation of disasters' impacts on cultural heritage and the protection of these assets (UNISDR, 2015).

Taking a different perspective, some authors recognize the intrinsic resilience of heritage premises as a key dimension for their durability (Holtorf, 2018; Seekamp & Jo, 2020). This is illustrated by their capacity to withstand disastrous situations while successfully adapting to new conditions. Such an approach was somehow previously implied in the Hyogo Framework for Action 2005–2015, in which the only reference to cultural heritage slightly appears in the third priority for action: "Use knowledge, innovation and education to build a culture of safety and resilience at all levels" (UNISDR, 2005, p. 6). Although the Sendai Framework reported the low prominence of this theme as a gap, it is quite interesting to note that the Hyogo Framework did not portray the cultural heritage as a fragile element. Instead, this framework depicted it more positively as a source of knowledge in terms of "disaster risks and protection options" (UNISDR, 2005, p. 9). In any case, the vulnerability of the built heritage, and not its resilience, has regularly been the focus of heritage organizations, reflecting a rather conservative approach toward the concerned assets.

Yet, this approach tends to imply the need for putting in place technological means, based on vernacular practices or state-of-the-art engineering devices, to overprotect the built heritage. Such interventionism may entail the threat of freezing historic monuments as an undesired drawback. It might ultimately constitute a moral hazard and underrate the role of the built heritage as a witness of "how people in the past have proven to be resilient and been capable of absorbing adversity in various ways" (Holtorf, 2018, p. 644). It might also compromise viewing the built heritage as a source of "learning from loss" that can favor its positive transformation (Seekamp & Jo, 2020, p. 42). This reasoning is somehow shared by Coetzee and Van Niekerk (2018), two DRR researchers not related to the heritage sphere who questioned more generally the pertinence of reducing all disaster risks. Based on the concept of "edge of chaos," they posited that "the elimination of risk" would also imply disregarding "learning and adaptation opportunities" that have been building the overall resilience of societies to face disasters across time (Coetzee & Van Niekerk, 2018, p. 478). Such a loss could arguably configure a cultural handicap. Eventually, policies for protecting heritage sites at any cost are thus economically and culturally unsustainable.

In this regard, it is worthwhile to recall that cultural heritage is a living legacy with material and immaterial dimensions, such as knowledge, beliefs, and value systems. These are fundamental aspects "that have a powerful influence on people's daily choices and behaviors" (UNISDR, 2013, p. 13). As such, the notion of integral adaptation, put forward by O'Brien and Hochachka (2011) in the context of climate change, seems to be a timely approach to be embraced by the heritage community. This concept regards adaptation as involving "not only adjustments in behaviors and systems, but also cognitive and cultural adaptations," as well as the "need to create adaptive strategies that are meaningful and make sense to people" (O'Brien & Hochachka, 2011, p. 98).

Through this conceptual standpoint, the material protection of heritage buildings and sites can therefore go hand in hand with the promotion of cultural resilience. Cornelius Holtorf, holder of the UNESCO Chair on Heritage Futures at Linnaeus University (Sweden), understands this as "the capability of a cultural system (consisting of cultural processes in relevant communities) to absorb adversity, deal with change and continue to develop" (2018, p. 639). Hence, when frozen by structural measures, heritage assets might fail to demonstrate cultural resilience. We explored elsewhere integral adaptation as a timely framework for supporting the incorporation of flood risk in contemporary urban projects through design (Hobeica & Hobeica, 2019). Nonetheless, it can also be useful to examine the interactions between

DRR interventions and the conservation of historic monuments in flood-prone sites, such as the Monastery of Santa Clara-a-Velha, an outstanding relic of the Portuguese late gothic.

3. Contextualizing the Monastery of Santa Clara-a-Velha's revitalization

Any consideration about the Monastery of Santa Clara-a-Velha must necessarily take into account the flood phenomenon in the basin of the Mondego, the largest entirely Portuguese river. Coimbra is settled in the middle reach of the Mondego, in a transition space marked by complex water dynamics. There, the river's profile changes from the upstream narrow valley, with steep slopes highly subject to erosion, to the downstream alluvial plain (Martins, 1951/1983). The Mondego River has a Mediterranean regime, presenting strong water-discharge variations throughout the year and recurrent winter floods. When more intense, these events used to jeopardize the agricultural fields downstream (Barbosa et al., 2006). For this reason, the river has been locally nicknamed "Basófias" (meaning mendacious or charlatan), a term that stresses its buffoon character and the idea that it could not be trusted (Cardielos et al., 2013).

Although floods have represented a problem mostly for the agricultural activities in the floodplain downstream of Coimbra, the city itself has not been totally spared from their effects. Indeed, the long-lasting battle against the waters in its center is illustrated by the abandonment or disappearance of several religious constructions on both riverbanks, being the Monastery of Santa Clara-a-Velha the most emblematic example (Martins, 1951/1983). Even if partially inundated for centuries, the Monastery's church has to some extent survived to testify the power of the water (Fig. 3.1).

Linked to the life of the Portuguese Queen Saint Isabel, the Monastery became a highly symbolic cultural icon of the city and indeed the nation, after having been gradually submerged due to recurrent floods (Bandeirinha, 2009). Following several attempts to adapt it to a wet environment and a series of resistance strategies (notably the building of new floor levels), the residing nuns ultimately retreated, moving in the 17th century to a new monastery built uphill (Côrte-Real, 2009). Since then, the Monastery's church had been adapted for housing and storage purposes, before being left abandoned. Its other premises were demolished for the reuse of their building materials elsewhere in the city.

Figure 3.1 The Mondego River besieging the church of the Monastery of Santa Clara-a-Velha (1955). Used with permission from Biblioteca Municipal de Coimbra.

4. The Monastery's revamping process

Governmental authorities have for long undertaken several resistance attempts, such as the ad hoc construction of dikes and landfills, to face floods in the lower segment of the Mondego basin. The diversion of the Mondego in the 18th century, carried out as the "correction of the river" to "defend" the agricultural fields in the floodplain, attenuated the problem (Martins, 1951/1983, p. 62). Yet, in less than a century, this solution proved to be insufficient, as the silting process subsequently favored the resurgence of significant floods, such as the ones experienced in 1900, 1915, 1924, and 1948 (Barbosa et al., 2006). This trend prompted the elaboration of a major structural intervention—the Mondego's hydraulic-exploitation plan—in 1962, combining several objectives at once: flood control, irrigation, water supply, and energy generation. Through the plan's implementation, the "full control" of the river was finally accomplished in the 1980s, with hard-regulation works comprising three dams upstream of Coimbra and, downstream of the city center, a dam bridge and river-training works (Sanches, 1996). These infrastructures have strongly reduced the frequency and magnitude of large floods around the city. Consequently, floods have been mentally transformed into an unlikely issue.

Also, the newly created *stable* waterscape encouraged the city's authorities to reconsider the river, not as a barrier anymore but as a connector between the two urban riverbanks. The regulation of the Mondego also triggered archaeological works in the ruins of the Monastery of Santa Clara-a-Velha, following a design competition in 1989 for the rehabilitation of its church. The related design brief called for ideas to improve the access to the church while keeping its ground floor as it had been for centuries: under water (Côrte-Real, 2009). Yet, the unexpected discovery of the cloister's ruins in very good conditions in 1995 led to the abandonment of the 1989 requalification project and the scaling-up of the archaeological works.

Following public debates and the assessment of alternatives to keep or not water within the site, the concerned institutions decided in 1998 to reinstate the Monastery's dry environment (Bandeirinha, 2009). Although being the most expensive choice, this option would enable the experience of the monument to its full potential. It would thus put an end to the seven-century-long "decadent enchantment" of the church (Bandeirinha, Olaio, & Correia, 2009, p. 3). Yet, the site is presently below the river's mean water level, and the water-pumping process used during the archaeological works had a prohibitive cost. Hence, the concerned authorities had to find a more enduring technique to maintain the envisaged dry condition. The adopted solution was the construction of a cofferdam around the Monastery's complex, conceived as a "genuine modern rampart that impedes the water from coming into the protected site" (Côrte-Real, 2009, p. 79).

The beginning of the cofferdam's works was accompanied by a particularly rainy period between December 2000 and January 2001, which led to a major flood on 27 January ("*unexpectedly,*" at least for two interviewees) (Fig. 3.2). In Coimbra's urban area, the discharge peak reached 1990 m^3/s, which is very close to the planned limit of the hydraulic system to protect the city. The left bank was the most impacted, although without severe damage. In contrast, the agricultural activities in the lower Mondego were strongly affected (Barbosa et al., 2006). This event shed light on a new type of human-related hazard closely linked to the uneven management of the Mondego's dams (Antunes do Carmo, 2014). One of the interviewed architects even presented this flood event as a "*human error*" that could have been simply avoided.

The January 2001 flood actually functioned as a reminder to all the involved stakeholders that a flood-prone site does not lose this key characteristic even when structural measures are in place. According to Louro and Lourenço (2005), despite being less exceptional than the 1948 event, the 2001 flood resulted in more material damage. This was mainly due to the fading of the flood-risk culture that had prevailed among the population living in the floodplain before the regulation works. Indeed, after the

Figure 3.2 In 2001, the river took hold of the Monastery and the archaeological site office (the white tents in the background). From Mateus, C., Cunha, L. (2013). A Oscilação do Atlântico Norte (NAO) e riscos climáticos em Coimbra, *Territorium, 20*, 42.

river-training interventions of the 1980s, the floodplain has been more intensively and not appropriately occupied. The higher exposure of more vulnerable assets thus contributed to the accrued losses (Louro & Lourenço, 2005).

Just after this flood event, the Portuguese national institution in charge of architectural heritage organized an international competition for a new requalification project for the Monastery's complex. The related competition brief revealed an ambiguous positioning as regards water. The project should highlight its presence in the site—keeping *"the memory of the probable threat"*—but should mask the cofferdam, required to ensure the new dry condition. As asserted by an interviewee, the cofferdam's layout cut the site transversally, somehow *deforming* the experience of the Monastery's complex as it was conceived. Despite being a critical structure for the project, the cofferdam was *"perhaps not valued enough by the heritage community,"* in the words of another interviewee. As argued by Bandeirinha (2009, p. 16),

> it is a pity that the cofferdam was then not considered an integral part of the future solution, but rather an abstract "technical" enclosing of the site (the latter being, on the other hand, understood as "cultural"). We now know, just as we knew at the time, that this opposition, regarded as anodyne, is only so in appearance. The separation between what is technical and what is cultural does not exist, and it is for this reason that the design of the cofferdam seems too detached from the site recovery plan, thus creating the need for a cosmetic treatment [...].

In contrast, the Portuguese practice Atelier 15 won the second competition with a proposal that, according to the same author, favored "lucidity, clarity, and temporal and cultural continuity" (Bandeirinha, 2009, p. 16).

5. The new premises of the Monastery

Apart from unveiling a renovated monument site, the Monastery of Santa Clara-a-Velha's revitalization shed light on some of the underlying sociocultural conditions regarding floods. These conditions somehow perpetuate the longstanding perception that floods in Coimbra are due to a river that refuses to be tamed—or whose control cannot be effective due to the uneven management of the defensive infrastructures in place. In the next two subsections, we examine both material and immaterial outcomes of the architectural intervention in the heritage site. Then we propose, in Section 5.3, a design and flood-management alternative that could have supported integral adaptation to floods.

5.1 Atelier 15's project for the Monastery

Finalized in 2002, Atelier 15's project contemplated the complete removal of mud and water, the rehabilitation of the Monastery's remains (the church and the adjacent cloister), and the construction of a museum to host the site's recent archaeological discoveries. In terms of dealing with floods, it included two complementary measures. The old church and the open-air ruins were protected by the cofferdam (Figs. 3.3 and 3.4), whereas the new institutional building, located outside the barrier (thus exposed to floods), was elevated on stilts (on the right, in Fig. 3.3).

As argued by one of the interviewed architects, the museum is built on stilts "*because we acknowledge that once every 100 years a flood can reach its ground-floor level. All the important materials that can be damaged during a flood were located upstairs.*" This formulation clearly expresses the acceptance of floods within the project's orientation. It also reveals that the common misinterpretation of the *centennial flood* notion is widespread even among well-informed building professionals. Yet, the same person specified: "*There is no guarantee that what was done will be completely effective [during a flood event],*" recognizing that the full control of floods is indeed unattainable. This condition was assumed since the earlier archaeological works. Located exactly where the museum's premises would be built, the temporary module used by the archaeologists (in the background of

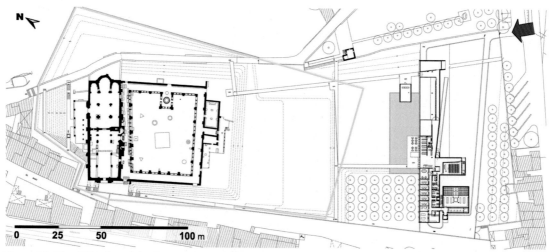

Figure 3.3 The 2002 plan of the Monastery's complex (in yellow, the cofferdam's layout; in blue, the reflective pool; the green arrow indicates the main entrance). Used with permission from Atelier 15 (adapted by the authors).

Fig. 3.2) had been placed "*over a mountain of stones.*" This strategy was a clear adaptation sign that, likewise, the design of the new building "*had to be proactive.*" Regarding the museum's premises, the interviewed archaeologist complemented:

> The building is a contemporary landmark, but one that is imbued with the history of a flood-prone territory. Anyone who stares at that space should see that fleeing from the water was not only a concern of the nuns who used to live there but also a present-day issue.

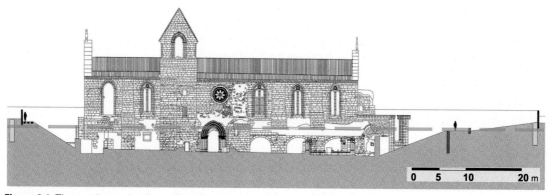

Figure 3.4 The southern elevation of the Monastery's church (in yellow, the cofferdam's location; in blue, the Mondego's mean water level). Used with permission from Atelier 15 (adapted by the authors).

The construction of the museum was qualified by an interviewee as a "*torment.*" The building of its foundations was particularly complex, because "*the site was always full of water,*" as the water table is very high. Moreover, on the several occasions that the pumps broke down, the water quickly percolated back into the site, delaying the works. Ultimately, by elevating the new building, the architects could introduce a reflecting pool near the cafeteria (indicated in blue in Fig. 3.3). This element emulates the mirror effect produced for centuries by the still water, which is also now constantly performed by the glass façade of the new building. The permanent exhibition in the museum also stresses the significance of water during the Monastery's lifetime, while artificial water sounds continuously emerge in the remains of the church.

Despite these contributions to flood awareness, the complex and the museum's exhibition completely overlook the cofferdam as the key device that made it possible to experience the Monastery in a dry environment. For instance, no clue is given to the visitors on the adopted solution, as if the water had simply disappeared on its own overnight. The cofferdam was not only physically masked through land grading but also veiled from people's perception and imagination. It should certainly not compete with the existing monument, but being confined to a technical and *neutral* barrier has prevented the cofferdam from being showcased as an early-21st-century engineering landmark. Moreover, masking such a device implied making flood risk less tangible. A more legible cofferdam could have brought to light both "*the memory of the probable threat*" and the contemporary existence of this still likely scenario.

Indeed, during more intense floods, the water could overtop the cofferdam (its highest point stands almost at the river's mean water level, 18 m). Thus, to prevent flood damage, the building contractor installed a system of reinforced-concrete walls and landfills above the cofferdam to reach the 20-m level (as indicated in Fig. 3.4, on the right). In any case, one can more accurately describe the cofferdam as a barrier exclusively related to floods involving percolation processes. Interestingly, it already exposed its adverse effects during the archaeological explorations. In 2002, a minor urban flood filled once more the site with water coming from the nearby hill. It was very difficult to dry up the area again, since the cofferdam "*functioned the other way round,*" impeding the water from naturally draining into the Mondego (Fig. 3.5).

As for the remains of the church, the project kept one of the openings in its upper floor, which recalls a former window that

Figure 3.5 A side effect of the cofferdam: water trapped after the 2002 pluvial flood. Used with permission from Atelier 15.

had functioned as the entrance to the building when its ground floor became continuously submerged. The project resorted to weathering steel (a water-resistant material) in the floor of the church, since "*it is still anticipated (or at least admitted) that more floods may happen.*" Labeled as "*ugly*" shortcomings of past floods by a stakeholder, the water marks on the church's walls (also shown in Fig. 3.5) inform the visitors about contingency and the passing of time. Indeed, in the words of the project architects, the church's ruins witness "the destructive power of time and the triumph of nature over culture" (Atelier 15, 2009, p. 14).

Such a remark recalls that despite the incorporation of a sophisticated technological solution, the heritage site certainly remains as susceptible to floods as it had always been. Accepting this certainty was one of the reasons behind the success of this project, which does not assume that floods have been simply relegated as an issue of the past. As formulated by the interviewed archaeologist: "*Although the site is now dry, [the imminence of the water's return] is really present: Turning off a switch is all it takes for the water to start rising; if we turn it off, two days later the area is filled with water again.*"

5.2 The intervention's sociocultural outcomes

The Monastery's history contains a lesson of contemporary relevance. Despite its proximity to the river, the site was probably not known to be floodable when the decision to occupy it was taken in the 14th century. However, mostly due to the river's silting process, exacerbated by the intensification of deforestation upstream, the context gradually changed after the location decision. The riverbed began to rise, and the river started to overflow into a wider area until finally reaching the Monastery's premises. Today, although protected by a cofferdam, the heritage site is not spared from having water inside it again. This lesson brings to the fore the pairs of opposites formed by certainty−permanence and unexpected−transient, as highlighted by an interviewee:

> Will we human beings be able to overcome nature with all these devices (dams etc.)? Well, it's a fact that we won't. The ruin is prepared, the new building is on stilts; as for the church, well, it has historically lived with the water, it may suffer some damage but this ought to be minimal.

Acceptance of the unforeseen was a departure point in the revamping process. Certainty here did not refer to idealized expected scenarios, but to the anticipation that, just like in the past, riverine floods will undoubtedly happen again—even if induced by different drivers. Moreover, all the interviewees recognized that the presence of water in the Monastery's complex had been fundamental for the very preservation of the artifacts now exposed in the museum and of the ruins themselves. In sum, *"the water has enabled the freezing of the heritage site,"* strongly contributing to its endurance.

Nonetheless, the two more recent significant flood events of January and February 2016 (Figs. 3.6 and 3.7) showed that the overall design could have better planned the functioning of the Monastery's premises during the periods when the site is again dominated by the water. The aforementioned trap effect was the main reason for the increased impacts of the two 2016 floods. During these events, the Mondego's waters were brought back to the site from the overloaded drainage system around the area, leaving the complex inoperative for almost three months. This backflow phenomenon, which could have been foreseen, was likely overlooked in the hydraulic modeling elaborated in the framework of the project. Together with the two pedestrian underpasses of the Inês de Castro Avenue (the shoreline road bordering the heritage complex, constructed on a dike in the 1950s), the setbacks of the overloaded drainage system are

Figure 3.6 The Mondego River besieging the entire Monastery complex in January 2016—a view of its main entrance. The authors (2016).

Figure 3.7 The Monastery as a living heritage complex—its ruins submerged again in 2016. The authors (2016).

presently the major factors behind the flooding of the Monastery's premises. Following a recent extensive DRR intervention—river-dredging works that lasted two years—Coimbra and the heritage complex were spared from the Mondego's December 2019 flood. Yet, this event heavily impacted the floodplain in the river's lower segment, indicating that flood risk and the ensuing damage were possibly transferred downstream.

The two 2016 floods did not reveal any degree of acceptance by the involved stakeholders. For instance, during the January 2016 event, the Monastery's administration alleged that it did not receive a timely warning, thus considering the management of the major dam upstream responsible for the damage incurred in the complex (Soldado, 2016). But one can reasonably argue that a warning would not have impeded the water's incursion into the site, as attested in February 2016. Although the uneven management of the dam may truly represent a new hazard, one cannot disregard the fact that the very existence of the Mondego's hydraulic system was the trigger for the revamping of the Monastery. Without these defensive structures in place, no heritage institution would have been enticed to undertake such an investment in the monument.

The material damage incurred from time to time due to floods could hence be considered a part of the price to be paid to ensure that the site is back to its dry state and accessible for people's enjoyment. The ensuing repair needs should thus be included as contingency costs, equating to regular operating expenditures or, since not on a yearly basis, to capital expenditures. This raises the question of whether the maintenance requirements related to floods were duly assessed before the commissioning of the new cultural facility. In this regard, one of the interviewed engineers noted:

> We know that [...] from time to time these natural conditions will happen, somehow damaging and deteriorating the prevailing situation and making it necessary to intervene. [...] And one can say that we can never relax, because these phenomena, which used to happen every 100 years, began to happen every 50 years and are now happening every 10 years. Nobody knows if in 20 years from now they will begin to happen every 2 years, right? The meteorological conditions that we were used to, the average values, and the statistics are being questioned daily by climate change. This worries us, it's a fact.

This assertion opens a debate related to the perception that the Mondego's floods are presently more frequent than expected. Would this be owing to epistemic risk—meaning that floods were not duly investigated at the outset—or to ontological risk—linked to the real dynamics of floods and their intrinsic changes? Or yet, can this perception be somehow related to the supposed *failure* of the hydraulic works that should *prevent* floods from happening at any cost? In the affirmative, this perception would again exemplify the fading of the flood-risk culture locally, as analyzed by Louro and Lourenço (2005), as well as the downsides of the quest for "eliminating" risk, as suggested by Coetzee and Van Niekerk (2018). The drawbacks of the defensive structures in place also include nurturing a false sense of stability that leads to the increase of moral-hazard situations.

The perspective taken in the project to evoke "the glorious past and the expiry of all things" (Atelier 15, 2009, p. 14) is well aligned with the cultural-resilience reasoning of Holtorf (2018). Yet, the resulting architectural intervention somehow reveals that a paradigm of stability underlay the design process. This inconsistency was patent, for instance, in the perception of the 2001 flood as an avoidable *"human error,"* implying that, as it happened once, it would not recur anymore. Viewing such events as potentially preventable human errors does not help to challenge the misconception that floods can be durably avoided by technological means (as somehow implicit in the Sendai Framework). Moreover, it does not favor the adoption of flood-adaptive stances. Thus, the intervention could also have more thoroughly embraced DRR if it had supported "enhancing human capacity to accept the possibility of loss over time" and "a continuous process of adaptation to shifting circumstances," as suggested by Holtorf (2018, p. 645). In other words, the material requalification of the heritage site could also have been performed as an opportunity to upgrade the local flood-risk culture to foster better living with future floods.

5.3 Welcoming floods through design

At this stage, we propose to envision an alternative design scenario: What if the intervention had pursued integral adaptation as a pillar and, therefore, had comprised adjustments in both physical and cultural dimensions? More precisely, what if the possibility of having water again in the site had been addressed through design? For instance, a contingent entrance to the church, to be used during and immediately after flood events, would even do justice to a key trait of heritage sites as put by the Venice Charter: "a monument is inseparable from the history to which it bears witness and from the setting in which it occurs" (ICOMOS, 1964). Yet, when asked about any envisaged alternative paths or movable installations that would make possible the visit of the Monastery when partially submerged—a question that probably sounded very naive before the 2016 floods, but pertinent afterward—the interviewed archaeologist replied in the following terms:

When there is a flood, the whole country is focused on the flood, not on the visitors of this Monastery or any other heritage site. We hope these situations to be very very sporadic [...]; what we have to do is to reorganize the space and make it functional again. The use of the space is compromised; it's a difficult coexistence in times of disaster.

Such an answer anticipated a paradox. Despite being aware of the potential presence of the water again in the complex and having prepared it to withstand its effects (Fig. 3.8), the involved stakeholders somehow dismissed the vicissitudes related to the occurrence of floods, namely their duration. As illustrated in 2002 and 2016, the water can remain trapped within the cofferdam for weeks. This fact not only affects the functioning and the financial viability of the museum but also hinders the city's tourism activity, an important sector of Coimbra's economy. For cities holding the UNESCO's World Heritage status, like Coimbra, the impacts on tourism are particularly significant among the indirect economic losses associated with disasters (UNDRR, 2019). In any case, the *unexpected* return of the water into the site did not need to represent a proper *disaster* in terms of indirect losses, had this scenario been adequately considered during the planning stage. If the unintentional trap effect is viewed through a positive prism, the cofferdam could also be praised for allowing the possibility of experiencing the complex under water, as it had been for centuries. Tourists and inhabitants alike would straightforwardly recognize the historic value of this wet scenario.

Thus, the project could have explored and assessed the possibility of *welcoming floods* (Hobeica & Santos, 2016). Devising a

Figure 3.8 The stilts fulfilling their protective role during the January 2016 flood. The authors (2016).

contingency plan that could make feasible the visits of the Monastery's submerged premises would certainly not be simple from the logistic point of view. Yet, the design could have kept the possibility of accessing the church differently, reviving the improvised entrance to the church in its previous uses. Such an alternative plan could provide a privileged, unexpected, and fascinating opportunity to temporarily experience entering the flooded church through the upper floor, as its past users had done for centuries. For instance, by resorting to temporary installations, the former access through the open window in the upper floor could have been kept. More than a simple memory, it could be used when the church's ground floor and the cloister's ruins are under water. This possibility was even put into practice during the rehabilitation works, as shown in Fig. 3.5 (the improvised entrance to the church, during the civil works, is visible on the left).

Such a flexible design could have given continuity to history, brought back to life during infrequent short periods. It would have done justice to the water and its crucial contribution to the preservation of the monument across time. Of course, some damage and disturbance would be inevitable, but one needs to recognize that damage can also "serve as a memory" (Seekamp & Jo, 2020, p. 41). Instead of being only blamed, the Monastery's new flooding episodes could thus be taken as unique opportunities to temporarily enjoy the heritage site in its historic wet state. Although using a dry scenario, a very similar perspective was even embraced by an ephemeral initiative carried out in 2012 by the institution in charge of the monument, as described by the interviewed archaeologist:

> [In 2012], we came up with an idea (a costless one), which was to recreate the 19th-century episode of an archaeologist who once decided to go on an expedition into the interior of the church: Not without fear, he and two other persons took a boat and torches to navigate the dark waters under those arches. [...] So we replicated this scene at night (without water, of course), with a few torches, and it was a success that exceeded our expectations: 600 people participated!

This initiative indicates that *welcoming floods* in this particular situation—that is, having a contingency entrance to be used during and rightly after flood events—could indeed have been highly valued by the Monastery's visitors. But as exposing the cofferdam was in opposition to the imagetic-protection stance then pursued, such an unconventional opening would probably not have been attuned with the overall design ambitions and the prevailingly esthetic idea of heritage value. Yet, both decisions—masking the cofferdam and not having an alternative entrance to the church— appear to be at odds with the DRR goal of promoting cultural resilience.

6. Blending heritage's physical and sociocultural dimensions

The Sendai Framework's third priority draws attention to the key role of DRR investments in the mitigation of disaster losses. However, by tending to focus on technological means, it neglects the involved sociocultural DRR dimensions. Indeed, DRR invest- ment decisions go much beyond choosing a hazard-mitigation strategy from an array of available alternatives. In the case of heritage sites, such a narrow understanding, coupled with a pres- ervationist mindset, might imply lower degrees of risk prepared- ness and higher cultural-resilience losses. Our analyses of the Monastery of Santa Clara-a-Velha's requalification process sug- gest that although the project's stakeholders—the heritage com- munity, municipal officers, architects, and engineers—clearly dealt with risk mitigation, they seem to have underrated flood preparedness. On two consecutive occasions in 2016, the heritage complex demonstrated to lack a proper contingency plan that en- visions its functioning during and after flood events. For instance, the possibility of visiting the Monastery while submerged would recognize the presence of the water as a legitimate historic layer. Without detracting from the interesting parts of the monument and without putting at risk its overall protection, this alternative could thus combine risk awareness and both spatial and cultural adaptations to flood (Hobeica & Hobeica, 2018).

In this case, the core structural measures in place (the cofferdam and the building's stilts) could indeed have gone hand in hand with more proactive and positive response and re- covery in actual flood situations. Such a combination would not prevent water from occasionally returning to the site. Yet, it could avoid the further decoupling between the site and the wa- ter, as illustrated by the 2016 floods' ensuing losses of opportunities—in cultural (sensitization) and business (continu- ity of operations) terms—and search for scapegoats. As the proj- ect did not foresee that with the climate-change prospects the site will probably be flooded more frequently, it also lost an op- portunity to promote a mindset adaptation to floods (Hobeica & Hobeica, 2019).

Among the lessons learned through this case that could be extrapolated to the conservation of heritage sites, the most important one is perhaps the need to holistically integrate DRR into their management. This would allow not only all the stages of the disaster cycle (including emergency and recovery) but also the cultural dimension of resilience to be duly taken into account in the design process. For instance, welcoming floods in the Santa Clara-a-Velha heritage site could have represented a ludic and pedagogical possibility to explore interconnections between history and memory. Indeed, accepting flood risk implies expecting wetter scenarios and exploring creative means to include this contingency into the design and management of heritage sites under revitalization processes.

Acknowledgments

This study was partially funded by the Portuguese Foundation for Science and Technology, through a PhD grant. The authors are thankful to the stakeholders who generously provided information to this research.

References

Antunes do Carmo, J. S. (2014). Environmental impacts of human action in watercourses. *Natural Hazards and Earth System Sciences, 2*(10), 6499–6530. https://doi.org/10.5194/nhessd-2-6499-2014

Atelier 15. (2009). Valorização do Mosteiro de Santa Clara-a-Velha em Coimbra [Revamping of the Monastery of Santa Clara-a-Velha in Coimbra]. *ECDJ, 12,* 6–15 (in Portuguese).

Bandeirinha, J. A. (2009). O que acontece quando a alma não é pequena: Ressurgimento de Santa Clara-a-Velha [What happens when the soul is not small: Resurgence of Santa Clara-a-Velha]. *Rua Larga, 25,* 14–19 (in Portuguese).

Bandeirinha, J. A., Olaio, A., & Correia, N. (2009). Editorial. *ECDJ, 12,* 3 (in Portuguese).

Barbosa, A. E., Alves, E., Cortes, R., Silva-Santos, P. M., Aguiar, F., & Ferreira, T. (2006). Evaluation of environmental impacts resulting from river regulation works: A case study from Portugal. In R. Ferreira, E. Alves, J. Leal, & A. Cardoso (Eds.), *Proceedings of the International Conference on Fluvial Hydraulics—River Flow 2006* (pp. 2081–2091). London: Taylor & Francis.

Cardielos, J. P., Lobo, R., Peixoto, P., Mota, E., Duxbury, N., & Caiado, P. (2013). Mondego: The dull murmur of the river. In R. Farinella (Org.), *Acqua come patrimonio: Esperienze e 'savoir faire' nella riqualificazione della città d'acqua e dei paesaggi fluviali* (pp. 106–125). Roma: Aracne.

Choay, F. (1992). *L'allégorie du patrimoine [The invention of the historic monument].* Paris: Seuil (in French).

Coetzee, C., & Van Niekerk, D. (2018). Should all disaster risks be reduced? A perspective from the systems concept of the edge of chaos. *Environmental Hazards, 17*(5), 470–481. https://doi.org/10.1080/17477891.2018.1463912

Côrte-Real, A. (2009). *Mosteiro de Santa Clara de Coimbra: Do convento à ruína, da ruína à contemporaneidade [Monastery of Santa Clara in Coimbra: From the convent to the ruins, from the ruins to contemporaneity]*. Coimbra: DRCC (in Portuguese).

Dawson, J. (2010). Thick description. In A. Mills, G. Durepos, & E. Wiebe (Eds.), *Encyclopedia of case study research* (pp. 942–944). Thousand Oaks, CA: Sage.

Foqué, R. (2010). *Building knowledge in architecture*. Brussels: UPA.

Hobeica, L., & Hobeica, A. (2018). Flood-prone urban parks: Linking flood adaptation and risk awareness. *Territorium, 25*(II), 143–160. https://doi.org/10.14195/1647-7723_25-2_12

Hobeica, L., & Hobeica, A. (2019). How adapted are built-environment professionals to flood adaptation? *International Journal of Disaster Resilience in the Built Environment, 10*(4), 248–259. https://doi.org/10.1108/IJDRBE-06-2019-0029

Hobeica, L., & Santos, P. (2016). Design with floods: From defence against a 'natural' threat to adaptation to a human-natural process. *International Journal of Safety and Security Engineering, 6*(3), 616–626. https://doi.org/10.2495/SAFE-V6-N3-616-626

Holtorf, C. (2018). Embracing change: How cultural resilience is increased through cultural heritage. *World Archaeology, 50*(4), 639–650. https://doi.org/10.1080/00438243.2018.1510340

ICOMOS (International Council on Monuments and Sites). (1964). *International Charter for the Conservation and Restoration of Monuments and Sites (The Venice Charter)*. Retrieved from https://www.icomos.org/charters/venice_e.pdf.

ICOMOS-ICORP (International Committee on Risk Preparedness of the International Council on Monuments and Sites). (May 23, 2013). *ICOMOS-ICORP Statement Heritage and Resilience*. Statement presented at the 4th Session of the Global Platform for Disaster Risk Reduction, Geneva. Retrieved from http://www.preventionweb.net/files/globalplatform/519de2efb7336 ICOMOS-ICORP_Statement_for_Global_Platform_-_23_May_2013_Final_Version.pdf.

Jigyasu, R. (2016). Reducing disaster risks to urban cultural heritage. *Journal of Heritage Management, 1*(1), 59–67.

Louro, S., & Lourenço, L. (2005). O comportamento hidrológico do Rio Mondego perante os valores de precipitação intensa, em Coimbra [The hydrological behavior of the Mondego River in the presence of intense precipitation values in Coimbra]. *Territorium, 12*, 19–27 (in Portuguese).

Martins, A. F. (1951/1983). Esta Coimbra… Alguns apontamentos para uma palestra [Coimbra… Some notes for a lecture]. *Cadernos de Geografia, 1*, 35–78 (in Portuguese).

O'Brien, K., & Hochachka, G. (2011). Integral adaptation to climate change. *Journal of Integral Theory and Practice, 5*(1), 89–102.

Sanches, R. (1996). *O problema secular do Mondego e a sua resolução [The protracted problem of the Mondego River and its solution]*. Lisbon: LNEC (in Portuguese).

Seekamp, E., & Jo, E. (2020). Resilience and transformation of heritage sites to accommodate for loss and learning in a changing climate. *Climatic Change, 162*, 41–55. https://doi.org/10.1007/s10584-020-02812-4

Soldado, C. (February 15, 2016). Prejuízo no Mosteiro de Santa Clara-a-Velha pode atingir os 600 mil euros [Damage to the Monastery of Santa Clara-a-Velha can reach 600,000 euros]. *Público*. Retrieved from https://www.

publico.pt./culturaipsilon/noticia/prejuizo-no-mosteiro-de-santa-claraavelha-pode-atingir-os-600-mil-euros-1723161?page=-1 (in Portuguese)

UNDRR (United Nations Office for Disaster Risk Reduction). (2019). *Global Assessment Report on Disaster Risk Reduction*. Retrieved from https://gar. unisdr.org.

UNISDR (United Nations Office for Disaster Risk Reduction). (2005). *Hyogo Framework for Action 2005–2015: Building the resilience of nations and communities to disasters*. Retrieved from https://www.unisdr.org/files/1037_hyogoframeworkforactionenglish.pdf.

UNISDR (United Nations Office for Disaster Risk Reduction). (May 13–23, 2013). *Heritage and resilience: Issues and opportunities for reducing disaster risk*. Background paper presented at the 4th Session of the Global Platform for Disaster Risk Reduction, Geneva. Retrieved from http://icorp.icomos.org/wp-content/uploads/2017/10/Heritage_and_Resilience_Report_for_UNISDR_2013.pdf.

UNISDR (United Nations Office for Disaster Risk Reduction). (2015). *Sendai Framework for Disaster Risk Reduction 2015–2030*. Retrieved from https://www.preventionweb.net/files/43291_sendaiframeworkfordrren.pdf.

Yin, R. (2009). *Case study research: Design and methods* (4th ed.). Thousand Oaks, CA: Sage.

B

Investing in new methods for resilience

Physical Services Index for flooding hazards

Charlotte Kendra Gotangco and Jairus Carmela Josol

Department of Environmental Science, School of Science and Engineering, and Ateneo Institute of Sustainability, Ateneo de Manila University, Quezon City, Metro Manila, Philippines

1. Understanding cascading hazards during urban-flooding events

Since the 1980s, climate- and weather-related hazards have been triggering much of the increase in disaster numbers and losses (Solecki, Leichenko, & O'Brien, 2011). Despite the numerous efforts to adapt to climate change and reduce disaster risks, the persistence of these disasters merits a reexamination of how we understand and manage them. It is important to note that the occurrence of climate- and weather-related events per se does not necessarily lead to disasters. Disasters only materialize when vulnerable populations and assets are subjected to and severely affected by these events. Using the framework of the United Nations Office for Disaster Risk Reduction (UNDRR), a disaster risk exists when hazard, exposure, and vulnerability come together (UNDRR, 2021a).

For the hazard component, many assessments tend to deal with rapid-onset phenomena discretely, such as determining the return period of extreme rainfall and flooding events. Although useful, the emphasis on these aspects alone might overlook how hazards interact with the natural and built environments. Such interaction may either modify how the hazards are manifested, or even trigger a cascade of secondary and tertiary hazards (Cutter, 2018). For example, the overall hazard resulting from extreme rainfall will be different if it occurs in a heavily built area compared with an area with more permeable surfaces. The first case would result in higher flood volumes. If the city also has waterways clogged with solid waste and contaminated with untreated human waste, then the contamination of floodwater can likewise contribute to diseases such as acute gastroenteritis or leptospirosis. From the perspective of human security and health, inadequate urban services, often construed as sources of vulnerability, can also be sources of harm for the exposed populations.

The cascading of secondary and tertiary hazards represents a class of multihazards that is easily overlooked. This is because these are not discrete events but arise from harmful conditions that exist in society. The UNDRR definition of multihazards refers to "the specific contexts where hazardous events may occur simultaneously, cascadingly, or cumulatively over time, and taking into account the potential interrelated effects" (UNDRR, 2021b, para. 4). The environmental and health hazards triggered by rainfall events are precisely conditions that persist—or are even aggravated—over time as populations grow and development goals are not met. From an entitlements perspective (Sen, 1981), it can be argued that the chronic underprovision of urban services and the deprivation of basic needs can constitute hazards themselves because they limit the potential for human flourishing.

The Sendai Framework for Disaster Risk Reduction for 2015–2030 explicitly articulated that approaches to disaster risk reduction (DRR) must be multihazard and multisectoral to be effective (UNISDR, 2015). Thus, for hazard assessments to be more comprehensive, they need to consider all the relevant anthropogenic and natural factors that can cause harm within any given place (Kappes, 2011; Kappes, Keiler, von Elverfeldt, & Glade, 2012). They should also take into account the potential relationships among these factors (Wang, He, & Weng, 2020). This includes identifying the interconnections among the different sectors responsible for urban services that affect hazard evolution. Combined with the vulnerabilities of the exposed populations and assets, this confluence of hazards may result in more disaster losses than the primary triggering event alone would have caused. The interventions to improve hazard management, therefore, should adopt a more holistic approach by considering the roles and interactions of climate and the features of the natural and built environments in shaping how multihazards develop.

Such a multihazard approach in an urban service context is consistent with, and perhaps provides a deeper perspective to, the understanding of the third priority of the Sendai Framework on investing in DRR for resilience. This priority recognizes that communities need to plan, fund, and implement both structural and nonstructural measures to enhance their economic, social, cultural, and health resilience. This entails the mainstreaming of disaster risk assessment and management into urban planning. In many contexts, investments made toward more general development goals, such as the provision of basic services to communities while preserving ecosystem functions, are also geared toward resilience in more than one sense. Addressing the gaps in these urban services, e.g., in the water and waste-management sectors, would result in reducing the cascade of secondary hazards during flooding events. Better services would also lead to healthier living conditions, which would reduce vulnerabilities in the community. This is a clear example of co-benefits, by which investments in quality of life and sustainability also contribute toward long-term resilience.

To better articulate and examine this class of multihazards, this chapter operationalizes the concept of cascading hazards for flooding events in an urban environment. We propose a Physical Services Index (PSI) that tracks the contributions of urban development to flooding hazards and chronic deprivation of human needs. Using Metro Manila, the Philippines, as a test case, a system-dynamics model calculates the PSI and simulates trends to explore the effects of population growth and urbanization. This chapter contributes to the analysis of complex urban dynamics through the integration of the different but interrelated hazards occurring during a flooding event, which will in turn help understand the factors and processes that shape these cascading hazards. The results can also inform the crafting of policies that are appropriate for urban development and that are co-beneficial for urban resilience.

2. An integrated flood-hazard approach to account for the effects of urban development on hazard evolution

This research implemented a system-dynamics approach to the study of interrelated and cascading hazards. The "system" aspect of the model emphasizes the interactions among sectors. The "dynamics" aspect enhances the analysis of hazards within a given place with their potential evolution over time, depending on the development choices made. System dynamics is the study of a system's structure to understand its behavior. Its salient feature is the use of feedback processes to understand how system components and actions have influenced current behavior, then to identify leverage points to modify future trends (Ahmad & Simonovic, 2000).

Systems approaches have been used to frame the concept of resilience (Helfgott, 2018) and have been recommended for decision-making on sustainability and resilience (Fiksel, 2006). System-dynamics modeling has been applied to complex structures and issues relevant to urbanization, such as hydrological systems and water resources (Khan, Yufeng, & Ahmad, 2009; Madani & Mariño, 2009; Stave, 2003), flood-reservoir management (Ahmad & Simonovic, 2000), energy consumption (Feng, Chen, & Zhang, 2013), transport and air quality (Armah, Yawson, & Pappoe, 2010; Chen, Ho, & Jan 2006; Sayyadi & Awasthi, 2020; Vafa-Arani, Jahani, Dashti, Heydari, & Moazen, 2014), and epidemics (Ritchie-Dunham & Méndez Galván, 1999; Sy et al., 2020). System dynamics has also been used to explore resource consumption and environmental footprints (Dacko, 2010; Jin, Xu, & Yang, 2009), sustainable supply-chain management (Rebs, Brandenburg, & Seuring, 2019; Saavedra, Fontes, & Freires, 2018), and socioecological systems as a whole (Elsawah et al., 2017). These studies help gain insights on how systems respond to different conditions, and therefore can serve as guides on how they will be managed.

Following a systems mindset, we conceptualized an integrated flood-hazard framework (Fig. 4.1) to account for urban areas' characteristics regarding both chronic hazards threatening human welfare (Kjellstrom et al., 2007) and sources of vulnerability that amplify flood hazards (Pescaroli & Alexander, 2015). These characteristics include land-cover changes and the state of urban services. The framework was the basis for developing the PSI and translating it into a causal-loop diagram (CLD) and a stock-and-flow model using a system-dynamics platform.

Based on the framework, there are five direct inputs to the PSI: water accessibility, water availability, sanitation and sewerage, municipal solid waste, and pollution loading and water quality. There are also two indirect influencing factors: urban growth and water balance. The scope of *urban growth* in this study focuses on the physical aspects of urbanization and how population growth drives changes in land cover and demand for basic services. Changes in land cover also affect the *water balance* by influencing run-off and groundwater recharge. In general, increases in built-up spaces lead to greater run-off and lesser recharge as they reduce the area available for infiltration.

From the raw water available, *water accessibility* refers to how water is extracted and distributed to meet the demand. *Water availability* represents the efficient use and sustainability of the surface-water and groundwater supplies. After the water is distributed and consumed, the state of *sanitation and sewerage* determines how wastewater is treated and disposed of. Because

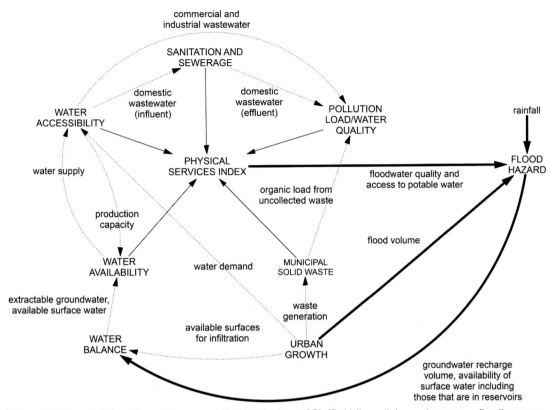

commercial and
industrial wastewater

SANITATION AND
SEWERAGE

domestic
wastewater
(influent)

domestic
wastewater
(effluent)

WATER
ACCESSIBILITY

POLLUTION
LOAD/WATER
QUALITY

rainfall

water supply

PHYSICAL
SERVICES INDEX

FLOOD
HAZARD

floodwater quality and
access to potable water

organic load from
uncollected waste

production
capacity

flood volume

WATER
AVAILABILITY

water demand

MUNICIPAL
SOLID WASTE

extractable groundwater,
available surface water

waste
generation

available surfaces
for infiltration

WATER
BALANCE

URBAN
GROWTH

groundwater recharge
volume, availability of
surface water including
those that are in reservoirs

Figure 4.1 Integrated flood-hazard framework for developing a PSI. (Bold lines: linkages between a flooding event and the state of the physical services). The authors.

flooding also triggers potential health hazards, we also need to track the domestic access to different types of sanitation facilities, which provide varying levels of protection from potential exposure to human waste. Like *sanitation and sewerage*, the ability to manage *municipal solid waste* from collection and disposal has the potential to both affect human health and contribute to pollution. Uncollected wastes and those that are disposed of in unmanaged dumpsites are possible breeding sites of disease vectors. They may also end up in waterways and surface waters, reducing their conveyance capacity and increasing their organic load, respectively. Thus, the ability to manage *pollution loading and water quality* of wastewater is also important in determining the extent of contamination of water bodies and in assessing whether ingestion and contact during a flood can pose health risks. The aggregated state of these physical services then determines the characteristics of the flood and the cascading hazards that can be experienced in the community.

3. Operationalizing the framework in Metro Manila

Metro Manila, as the capital of the Philippines, is the national center of socioeconomic and political activities. It has a land area of 636 km^2 and a population of 12.9 million people in 2015, which corresponds to 20,785 persons per km^2 (PSA, 2016). Though accounting for only 0.2% of the country's land area, Metro Manila contributed to 36% of the total GDP in 2018 (NEDA, 2018). Metro Manila is composed of 17 political subunits that function independently. Thus, planning is often fragmented, leading to a host of problems such as chronic water shortage, improperly disposed-of solid waste, and inadequate sanitation facilities. Owing to the area's topography, flooding is also a persistent problem especially during the rainy season (Mercado, Kawamura, & Amaguchi, 2020). Traditionally, each of these problems has been managed by different entities or regulatory bodies separately, with little integration.

We applied our integrated hazard framework in Metro Manila to assess whether a systems approach could provide better insights on how these multiple but interrelated problems can be managed. We constructed a CLD to identify the specific components that shape the hazards and to qualitatively assess their interrelationships (Fig. 4.2). The variables in the CLD represent

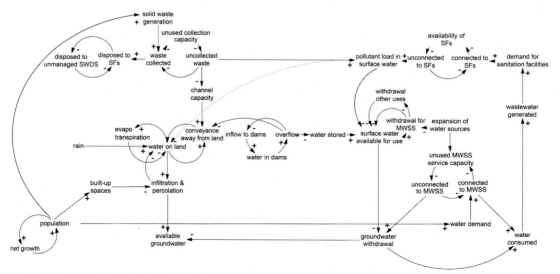

Figure 4.2 A causal-loop diagram of the short- and long-term processes that affect how hazards develop in Metro Manila, based on the components of the integrated hazard framework. (MWSS, which stands for Metropolitan Waterworks and Sewerage System, is the government agency responsible for supplying the water needs of the metropolis through private concessionaires. SWDS and SFs stand for solid-waste disposal sites and sanitary landfills, respectively.) The authors.

the physical-service sectors in the integrated framework (i.e., the variables pertaining to solid waste, sanitation, water, and urban growth, and the relationships among them). The (+) and (−) signs refer to the directions of change: (+) when an increase/decrease in one variable also leads to an increase/decrease in another, and (−) when the variables change in inverse directions, an increase in one leading to a decrease in the other. The linkages, however, do not detail the relationships deterministically but are aggregated according to their ability to contribute to the variable of interest. We proposed that this level of detail should suffice because the purpose of the model was not to achieve accurate forecasts but to provide tractable indicators. From these, we could gain insights on the interaction of different hazard components and the mutual conditioning of chronic hazards and short-term hazard events.

We then translated the CLD into systems notation (in stocks and flows) using *Vensim®* Personal Learning Edition. We organized the stock-and-flow model into modules based on the direct and indirect inputs to the PSI in the integrated framework. From surveys of available data and a meta-analysis of the literature, we parameterized the model using the conditions of Metro Manila.

3.1 Urban growth

From 1980 to 2010, the population of Metro Manila grew at an average rate of 2% per year and is projected to grow by 1% per year in the coming decades (PSA, 2016). The proportion of unbuilt and forested land declined to give way to residential and commercial spaces. Using this information, we derived a regression function to simulate how changes in population affect land cover. We then used the information on land cover to derive an area-wide run-off coefficient for Metro Manila. We used both the population and run-off coefficients as inputs to the other modules of the model. The population in the urban-growth component drives water demand, particularly domestic demand.

3.2 Water balance

Rainfall in the Philippines exhibits strong seasonality and high spatial variability (PAGASA, 2018). In the Metro Manila area, an increasing trend in rainfall was observed in the September, October, and November months from 1950 to 2010, whereas the trends in other seasons were not statistically significant (PAGASA, 2018). Projections of future rainfall vary as well. The midrange

(A1B) scenario of the Intergovernmental Panel on Climate Change forecasts slightly wetter conditions in Metro Manila in both the near term (until 2035) and medium term (2036–2065) (IPCC, 2000), compared with the 1970–2000 baseline (PAGASA, 2011). However, both the mid- and high-emissions scenarios (RCP 4.5 and 8.5, respectively) of the new set of Representative Concentration Pathways (RCP) (Van Vuuren et al., 2011) reflect slightly drier conditions when considering the median amounts modeled for Metro Manila by the mid-21st century (PAGASA, 2018).

We used the rainfall projections under the A1B scenario in this study for continuity (there are no near-term projections in the more recent study employing the RCP). This module estimated how such changes in rainfall patterns along with the changes in land cover could affect local water balance (Eq. 4.1). Infiltration (I), which is considered as the upper limit of groundwater recharge, is simply precipitation (P) net of run-off (R) and evapotranspiration (E).

$$I = P - R - E \qquad (4.1)$$

3.3 Water availability and water accessibility

Approximately 97% of the water needs in Metro Manila are derived from surface water from a multipurpose reservoir in an adjacent province. The amount of water allocated to Metro Manila is balanced with other competing uses of the reservoir such as irrigation, electricity generation, and environmental use. From the total allocation, only a fraction of the water reaches the consumer as 39% to 65% is lost during distribution (JICA & NWRB, 2003). The bulk of the surface water is consumed for domestic use, followed by commercial and industrial uses. The remaining water needs of the metropolis are met by extracting groundwater whose recharge is estimated in the water-balance module.

The *water availability* component of the model examined the sustainability of the resource by computing a groundwater index and a surface-water index. For groundwater, the index is simply the balance between extraction and recharge. For surface water, the index is the production efficiency relative to the amount of allocation. The *water balance* affects groundwater output by mediating the availability of extractable groundwater.

The *water accessibility* module simulated the dynamics of supply and demand. On the supply side, we estimated the surface water from a rainfall-reservoir allocation function subtracted by the distribution losses, and the groundwater from government-issued extraction permits. On the demand side, we estimated the levels of commercial, industrial, and domestic-water needs from population and affluence.

3.4 Sanitation and sewerage

Whereas sanitation and sewerage encompass the entire process of managing human waste, this module adopted a narrower definition and referred to the physical infrastructures used to contain and treat human waste. In Metro Manila, 94% of the population has access to some form of sanitation facility, 71% of which being to toilets with septic tanks (MWSS, 2005). However, only 20% of toilets with septic tanks are connected to treatment facilities or are regularly desludged (JICA & MWSS, 2009). These data indicate that most sanitation facilities discharge virtually untreated waste into the storm drains and waterways.

From a systems' perspective of hydrological-hazard development, effective sanitation facilities are crucial in two aspects. First, in preventing direct exposure to raw sewage that can affect human health, and second, in treating human waste that can affect the quality of the receiving water body. To account for this dual role, this module simulated the proportion of the population with access to sanitation facilities that afford different levels of protection from and treatment of human waste. These comprise toilets with septic tanks that are regularly desludged or are connected to a centralized wastewater treatment plant, toilets with septic tanks that are not regularly desludged, toilets without septic tanks, and pit latrines.

3.5 Municipal solid waste

In Metro Manila, a person generates approximately 0.7 kg of garbage per day (NSWMC, 2015), consisting of 45% kitchen waste, 17% paper, 16% plastic, 7% garden waste, and 5% metal (JICA & MMDA, 1999). Of the total waste generated, only 6% is recycled and 63% goes to final disposal (JICA & MMDA, 1999). The remaining fraction is burned, composted, or dumped illegally, which could eventually end up in waterways and reduce their conveyance capacity and increase their organic content.

Like human waste, solid waste is a hazard both in its potential to affect human health and in its capacity to alter the movement and quality of the water. Uncollected wastes and those that are disposed of in unmanaged sites are possible breeding spots of disease vectors. They could likewise end up in waterways, potentially reducing their capacity to effectively drain water or increase the organic load of the receiving water body. Given these, the module simulated the waste streams from generation to final disposal. Because data on waste generation are not collected regularly, we used per capita energy consumption as a proxy (Bogner & Matthews, 2003). We considered the following modes of final disposal, each affording different levels of management and protection: recycled, disposed of in streams and waterways, disposed of in controlled facilities, disposed of in sanitary landfills, and disposed of elsewhere.

3.6 Pollution load and water quality

The poor state of physical services, particularly wastewater and solid-waste management, affects the quality of the receiving water bodies. In Metro Manila, major water bodies are heavily polluted in such a way that they are unfit for fisheries and even secondary-contact recreation. We calculated the pollution load of wastewater from its volume and average composition. We assumed that all the water that is consumed (in *Water accessibility*) eventually becomes wastewater. We estimated the average composition of domestic wastewater from the treatment efficiency of the sanitation facilities (in *Sanitation and sewerage*), whereas we approximated the commercial and industrial wastewater based on the level of compliance to the national effluent standards. For solid waste, we calculated organic load from the volume and composition of solid waste that ends up in the waterways using the first-order decay model (IPCC, 2019). We estimated the final concentration pollutants ($[A]$) in the water bodies using Eq. (4.2), in which AC refers to the assimilative capacity of the receiving waterbody, m to the mass of the pollutant, and Vw to the volume of wastewater, whereas *dom*, *ind*, *comm*, and *msw* refer to domestic, industrial, commercial, and municipal sources of waste, respectively. Pollutants could refer to the following: biological oxygen demand (BOD), chemical oxygen demand (COD), total suspended solids (TSS), and fecal coliform.

$$[A] = [(m_{domA} + m_{indA} + m_{commA} + m_{mswA}) / V_w] \times (1 - AC) \quad (4.2)$$

3.7 Aggregated Physical Services Index

Using the outputs of the preceding modules, we derived five individual indices corresponding to the five direct inputs in the integrated flood-hazard framework. These indices are essentially a ratio of the availability of the service or the resource and the demand. The indices are from zero to positive, with one as the equilibrium when the physical services are only meeting the demand. Moreover, values less than one indicate unmet demands or service deficits, whereas values exceeding one indicate supply exceeding demand or service surpluses. The system-dynamics model allowed for the simulation of these indices over time to measure performance as the urban population grows and/or available vacant land is converted into built-up surfaces.

Table 4.1 shows the pertinent equations describing how we computed each index. We then combined the five indices to derive the PSI (Eq. 4.3). For simplicity, the indices were assigned equal weights (Sharpe & Andrews, 2012).

$$PSI = \left(W_{av} + W_{acc} + W_{qual} + MSW + SS\right)/5 \qquad (4.3)$$

The index for water availability combines the sustainability of groundwater and of surface water according to the balance ratios described in Section 3.3. The water-accessibility index examines

---

Table 4.1 Equations for computing the indices of the integrated PSI.

Equations for indices of the PSI	Legend
$W_{av} = (I_{SW} + I_{GW})/2$	W_{av} = water availability I_{SW} = surface-water index I_{GW} = groundwater index
$W_{acc} = W_d/eW_s$	W_{acc} = water accessibility W_d = water demand (all sectors) eW_s = effective supply (groundwater and piped water)
$W_{qual} = STD_{WQPA}/C_{WQPA}$	W_{qual} = water quality STD_{WQPA} = standard concentration of parameter A C_{WQPA} = ambient concentration of parameter A
$SS = (PD_{SFA} + PS_{SFA})/2$	SS = sanitation and sewerage PD_{SFA} = protection from direct contact with human waste PS_{SFA} = capacity to neutralize waste SFA = type of sanitation facility
$MSW = (F_R + F_{SLF} + 0.5F_{CDF})/Vt_{MSW}$	MSW = municipal solid waste F_R = fraction recycled F_{SLF} = fraction to sanitary landfills F_{CDF} = fraction to controlled disposal facilities Vt_{MSW} = total volume of waste generated

whether water demands are met or not. Since the bulk of the water needs are supplied by MWSS, it serves as an indicator of the capability of MWSS and its concessionaires to deliver water to the population.

We compared the water-quality parameters BOD, COD, and TSS against the effluent standards appropriate for the rivers in Metro Manila. Fecal coliform was compared against the drinking-water standard because of the possibility of ingestion and contamination of drinking water during flood events. We computed for individual pollutant indices, then the final water-pollution-load index was simply the average of the individual indices. Different modules contributed to the calculation of pollutant concentrations. The volume of wastewater generated was derived from *water accessibility*, whereas pollutant load from domestic waste was estimated from the relative fractions of domestic wastewater disposed of in different sanitation facilities. Finally, organic load from solid waste was derived from the uncollected waste fraction in the municipal solid-waste component.

For the sanitation and sewerage index, we considered only the contribution to health hazards because the impact on water quality was already included in the pollution-load component. This index comprises two groups, direct and indirect effects, because sanitation facilities provide different levels of physical protection against acute and chronic exposure to human waste. For the direct subgroup, populations with access to any type of sanitation facility carried a value of one because these facilities prevent exposure to human waste. Populations without sanitation facilities were assigned a value of zero. The average of the two generates the index representing protection from the direct contact with human waste. Chronic exposure to partially treated waste can affect health (e.g., persistence of diarrhea and malnutrition). Thus, for the indirect subgroup, the different types of sanitation facilities were assigned values according to their capacity to neutralize waste. We assumed that populations connected to the sewerage system were completely protected and assigned a value of one. The run-off ratio of the sewerage system was then used to derive the protection factors of populations with other types of sanitation facilities. The sanitation and sewarage component index is the average of the two subindices, representing the two groups. The fraction of the population with access to different types of sanitation facilities is dependent on the relative growth of sanitation facilities vis-à-vis the one of the population. This component is therefore related to the urban-growth component.

In the case of municipal solid waste, aside from contributing to the organic load of surface waters, improperly disposed-of waste is a possible breeding spot for disease vectors. It can also clog the waterways and reduce their conveyance capacity. Improperly built and maintained disposal sites, collapse of waste piles, fires, and leachate contamination of nearby water bodies and groundwater constitute potential hazards. Taking these into account, we assigned protection factors to each management approach: 1 for recycled waste or waste in sanitary landfills; 0.5 for waste in controlled dumpsites; and 0 for illegally/improperly disposed-of waste. The municipal solid-waste index was then calculated by weighting the protection factors with the corresponding waste proportions.

4. Modeling the Physical Services Index of Metro Manila through system dynamics

We tested three scenarios to explore the effect of urbanization on the PSI: (1) no population growth but continued increase in the built-up surfaces; (2) population increases but built-up surfaces remain the same; and (3) increase in both population and built-up surfaces. Due to challenges with data availability, the data and trends we used for the simulations were sourced from reports produced at different years (we mentioned these reports in Section 3). We acknowledge that this may introduce a degree of inconsistency in the model. Yet, we contend that as the purpose of the system-dynamics model is to understand (rather than to forecast) behavior, the trends that it generated suffice to draw insights on the factors influencing the PSI. In this case, we compared the relative importance of population growth with the increase in built-up area.

For groundwater resources, the scenarios whose built-up surfaces remained at 2010 levels had higher recharge rates. However, because the recharge levels were markedly less than the withdrawal quotas, we observed groundwater mining in all scenarios. In terms of the balance between effective water supply and demand, the zero-population-growth scenarios had higher accessibility-index values resulting from lesser surface-water demand. For the availability of water resources, the scenarios with no increase in built-up surfaces scored higher (Fig. 4.3) because lower run-off potential allows for more groundwater recharge.

In terms of managing wastes, the total volume of waste generated increased with population growth, but we assumed that

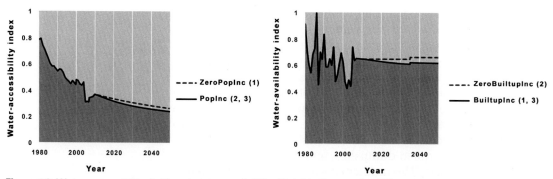

Figure 4.3 Water-accessibility (left) and water-availability (right) indices under different development scenarios. The authors.

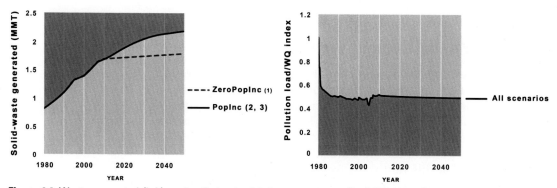

Figure 4.4 Waste generated (left), and pollution-load index or water quality (WQ) index (right, all values the same) under different development scenarios. The authors.

the relative proportions of people with access to different waste-management options would also grow with the population. Given such an assumption, the index scores related to environmental quality such as pollution load were similar under the different scenarios. However, even with the assumption that urban services grow proportionally with the population, the scores were still suboptimal (Fig. 4.4). This means that urban services need to outpace population growth if they are to fully meet the demand.

For the aggregated PSI, the scenario with an increasing population without the corresponding increase in built-up surfaces scored the highest, followed by the one with zero population growth with increasing built-up surfaces (Fig. 4.5). The scenario with increasing population and built-up surfaces (PopInc_BuiltupInc) scored the lowest. These findings demonstrate that if

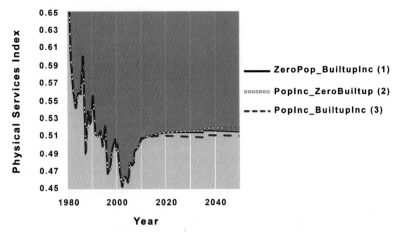

Figure 4.5 Integrated Physical Services Index under different development scenarios for the years 1980–2050 (up to 2010, the graph reflects primarily historical data). The authors.

groundwater extraction is regulated (i.e., through government enforcement of the permit system), and liquid- and solid-waste management services increase proportionally with the population, the effects of population growth on the PSI could be tempered by less urbanization. The converse is true for the scenario of zero population growth with continued urbanization (ZeroPop_BuiltupInc), whereby the gains due to demand reduction were outweighed by a reduction in resource availability (groundwater recharge). Note that, in Fig. 4.5, the small increase after the year 2035 is primarily due to an increase in annual rainfall in the projection centered around 2050, as explained in Section 3.2, and the corresponding effect on groundwater recharge and the water-availability index. However, had the RCP scenarios been employed, no change or a small decrease would have been observed, in view of slightly lower amounts of projected rainfall.

The graph in Fig. 4.5 includes the historical PSI to compare the past performance of physical services with the present and the potential future (if current trends are to continue without interventions). The index values indicate that deficits in the PSI and its indices are not a recent issue—Metro Manila has long been unable to meet the basic demands of its population. The PSI rapidly declined from 1980, and though there was some improvement after the year 2000, this has not been enough to sufficiently provide for the population. This means that during extreme rainfall events, these unmet urban services will interact with floodwaters in such a way that populations will need to cope with not only flood volumes but also other flooding hazards. These cascading flood hazards include limited clean-water availability and exposure to untreated wastewater and uncollected garbage.

Moreover, even without a trigger event and under a conservative scenario with no increases in population and built surfaces, the overall condition of the physical services remained suboptimal. Thus, typical DRR interventions focusing on limiting flood volumes, such as the construction of flood-control infrastructures and planning regulations that prevent further increases in built surfaces, will no longer suffice. Reducing the deficits in urban services will also be needed to address the potential cascading flood hazards. This is also sensible from a human security and health perspective because chronic deprivation is already a hazard in itself. Thus, the investments toward resilience that will be most effective are those that are co-beneficial, i.e., those that contribute simultaneously to basic development targets (such as the Sustainable Development Goals) and the Sendai Framework's priorities.

5. Assessing flood hazards holistically and optimizing resilience investments through the Physical Services Index

As a response to the narrow framing of flooding as a simple problem of excess water and the need to consider cascading hazards, we developed a Physical Services Index based on a more comprehensive and integrated flood-hazard framework. In particular, the PSI accounts for the twofold role of physical urban services in the amplification and cascading of flooding and in the identification of service and development gaps. Converted into a system-dynamics model, the PSI demonstrated its utility in assessing how different development configurations affect hazard processes and levels over time. Specifically, the findings show that the physical aspect of urbanization—the conversion of land into built-up surfaces—exerts a larger influence on the PSI than the population growth per se. The PSI is likewise useful for identifying tradeoffs across different modules as well as solutions that are robust and offer multiple benefits. The measures that seek only to reduce flood volume are inadequate. These need to be paired with interventions to improve urban services. Such initiatives would potentially meet both disaster-resilience and development objectives, and perhaps even climate-adaptation goals.

As we showed in this chapter, the primary contribution of the PSI is its conceptual expansion of "flooding hazards." In particular, by broadening the notion of flood hazards to include the role of physical urban services—not just in mediating the timing and volume of flooding, but also in considering the health hazards of floodwater—it widens the horizon of flood-management and resilience responses, often limited to flood-control infrastructures and measures to address acute shocks. This reframing opens the possibility of pursuing nonstructural interventions, including measures that address chronic stresses and those that may not be explicitly hazard related but are development oriented. As a composite index that measures the adequacy of basic services (or the hazardousness of chronic deprivation) and maps the interconnections of these services through system dynamics, the PSI is also useful in identifying priority areas. It is likewise useful in providing a transparent framework for addressing tradeoffs and detecting synergies across services. In Metro Manila, for example, flood-resilience measures might be more about satisfying unmet demands for basic services rather than simply making existing critical services hazard-proof and limiting floodwater.

Aligned with the larger context of the Sendai Framework, the PSI shows that addressing development gaps is a crucial first step toward pursuing risk-reduction priorities. Given the action-oriented promise of the Sendai Framework, the PSI helps operationalize this through its different modules that allow the exploration of possible pathways to inform actions and achieve targets. The PSI is able to support the third priority of the Sendai Framework toward investing in measures to reduce risk and enhance resilience. Investment decisions for resilience or hazard mitigation rarely occur in a vacuum but are made in the context of other objectives and priorities. At the global level, for example, the world's nations also committed to the Paris Agreement and the Sustainable Development Goals in 2015. One of the main challenges for the implementation of the Sendai Framework is promoting synergies with these other agendas. This involves both the conceptual task of articulating and understanding the linkages between resilience and sustainability (Marchese et al., 2018), and the practical undertaking of harmonizing the work of the many governmental entities involved in these two spheres. The PSI addresses the conceptual challenge more concretely by demonstrating how hazard processes are linked to development choices and climate change. Given the flooding in Metro Manila, for instance, the PSI demonstrated that investments in the quality of life and sustainability also contribute toward long-term resilience and climate adaptation. For the third priority of the Sendai Framework, the PSI helps identify opportunities for cost-effective investments that simultaneously address multiple objectives.

Acknowledgments

This work was partially supported by the following projects: Harmonizing FORIN for Climate Change Adaptation and Disaster Risk Management, funded by the International Council for Science (ICSU) and the US National Science Foundation; and the Capacity Building in Asia for Resilience Education (CABARET), co-funded by the Erasmus+ Program of the European Union. The European Commission's support for the production of this chapter does not constitute an endorsement of its contents, which only reflect the views of its authors, and the Commission cannot be held responsible for any use which may be made of the information contained therein.

References

Ahmad, S., & Simonovic, S. P. (2000). System dynamics modeling of reservoir operations for flood management. *Journal of Computing in Civil Engineering, 14*(July), 190−198. https://doi.org/10.1061/(ASCE)0887-3801(2000)14:3(190)

Armah, F. A., Yawson, D. O., & Pappoe, A. A. N. M. (2010). A systems dynamics approach to explore traffic congestion and air pollution link in the city of Accra, Ghana. *Sustainability, 2*(1), 252−265. https://doi.org/10.3390/su2010252

Bogner, J., & Matthews, E. (2003). Global methane emissions from landfills: New methodology and annual estimates 1980−1996. *Global Biogeochemical Cycles, 17*(2). https://doi.org/10.1029/2002gb001913

Chen, M., Ho, T., & Jan, C. (2006). A system dynamics model of sustainable urban development: Assessing air purification policies at Taipei City. *Asian Pacific Planning Review, 4*(1), 29−52. Retrieved from http://personal.its.ac.id/files/material/2955-cahyono_urplan-SustainableUrbanDevelopment.pdf.

Cutter, S. L. (2018). Compound, cascading, or complex disasters: What's in a name? *Environment: Science and Policy for Sustainable Development, 60*(6), 16−25. https://doi.org/10.1080/00139157.2018.1517518

Dacko, M. (2010). Systems dynamics in modeling sustainable management of the environment and its resources. *Polish Journal of Environmental Studies, 19*(4), 699−706.

Elsawah, S., Pierce, S. A., Hamilton, S. H., van Delden, H., Haase, D., Elmahdi, A., & Jakeman, A. J. (2017). An overview of the system dynamics process for integrated modelling of socio-ecological systems: Lessons on good modelling practice from five case studies. *Environmental Modelling and Software, 93*, 127−145. https://doi.org/10.1016/j.envsoft.2017.03.001

Feng, Y. Y., Chen, S. Q., & Zhang, L. X. (2013). System dynamics modeling for urban energy consumption and CO_2 emissions: A case study of Beijing, China. *Ecological Modelling, 252*(1), 44−52. https://doi.org/10.1016/j.ecolmodel.2012.09.008

Fiksel, J. (2006). Sustainability and resilience: Toward a systems approach. *Sustainability: Science, Practice and Policy, 2*(2), 14−21. https://doi.org/10.1080/15487733.2006.11907980

Helfgott, A. (2018). Operationalising systemic resilience. *European Journal of Operational Research, 268*(3), 852−864. https://doi.org/10.1016/j.ejor.2017.11.056

IPCC (Intergovernmental Panel on Climate Change). (2000). *IPCC special report—Emissions scenarios, summary for policymakers.* Retrieved from https://www.ipcc.ch/site/assets/uploads/2018/03/sres-en.pdf.

IPCC (Intergovernmental Panel on Climate Change). (2019). *2019 refinement to the 2006 IPCC guidelines for national greenhouse gas inventories.* Retrieved from https://www.ipcc-nggip.iges.or.jp/public/2019rf/index.html.

JICA (Japan International Cooperation Agency), & MMDA (Metropolitan Manila Development Authority). (1999). *The study on solid waste management for Metro Manila in the Republic of the Philippines. Final report, main report I (master plan).* Retrieved from https://openjicareport.jica.go.jp/pdf/11495603_01.pdf.

JICA (Japan International Cooperation Agency), & MWSS (Metropolitan Waterworks and Sewerage System). (2009). *Preparatory survey for Metro Manila sewerage and sanitation improvement: Final report.* Retrieved from https://openjicareport.jica.go.jp/pdf/11948882_01.pdf.

JICA (Japan International Cooperation Agency), & NWRB (National Water Resources Board). (2003). *The study on water resources development for Metro Manila in the Republic of the Philippines.* Retrieved from https://openjicareport.jica.go.jp/pdf/11721933_01.pdf.

Jin, W., Xu, L., & Yang, Z. (2009). Modeling a policy making framework for urban sustainability: Incorporating system dynamics into the ecological footprint. *Ecological Economics, 68*(12), 2938–2949. https://doi.org/10.1016/j.ecolecon.2009.06.010

Kappes, M. S. (2011). *Multi-hazard risk analyses: A concept and its implementation.* Doctoral thesis. Vienna, Austria: University of Vienna. Retrieved from http://othes.univie.ac.at/15973/1/2011-08-03_0848032.pdf.

Kappes, M. S., Keiler, M., von Elverfeldt, K., & Glade, T. (2012). Challenges of analyzing multi-hazard risk: A review. *Natural Hazards, 64*(2), 1925–1958. https://doi.org/10.1007/s11069-012-0294-2

Khan, S., Yufeng, L., & Ahmad, A. (2009). Analysing complex behaviour of hydrological systems through a system dynamics approach. *Environmental Modelling and Software, 24*(12), 1363–1372. https://doi.org/10.1016/j.envsoft.2007.06.006

Kjellstrom, T., Friel, S., Dixon, J., Corvalan, C., Rehfuess, E., Campbell-Lendrum, D., … Bartram, J. (2007). Urban environmental health hazards and health equity. *Journal of Urban Health, 84*(Suppl. 1), 86–97. https://doi.org/10.1007/s11524-007-9171-9

Madani, K., & Mariño, M. A. (2009). System dynamics analysis for managing Iran's Zayandeh-Rud River basin. *Water Resources Management, 23*(11), 2163–2187. https://doi.org/10.1007/s11269-008-9376-z

Marchese, D., Reynolds, E., Bates, M. E., Morgan, H., Clark, S. S., & Linkov, I. (2018). Resilience and sustainability: Similarities and differences in environmental management applications. *Science of the Total Environment, 613*, 1275–1283. https://doi.org/10.1016/j.scitotenv.2017.09.086

Mercado, J. M., Kawamura, A., & Amaguchi, H. (2020). Interrelationships of the barriers to integrated flood risk management adaptation in Metro Manila, Philippines. *International Journal of Disaster Risk Reduction, 49*, 1–12. https://doi.org/10.1016/j.ijdrr.2020.101683

MWSS (Metropolitan Waterworks and Sewerage System). (2005). *Water supply, sewerage and sanitation master plan for Metro Manila. Final report.* Retrieved from https://openjicareport.jica.go.jp/pdf/11948882_07.pdf.

NEDA (National Economic Development Authority). (2018). *Socio-economic report.* Retrieved from https://www.neda.gov.ph/wp-content/uploads/2019/09/SER-2018_Posted-Chapters.pdf.

NSWMC (National Solid Waste Management Commission). (2015). *National solid waste management status report (2008-2014)*. Retrieved from https://nswmc.emb.gov.ph/wp-content/uploads/2016/06/Solid-Wastefinaldraft-12.29.15.pdf.

PAGASA (Philippine Atmospheric, Geophysical and Astronomical Services Administration). (2011). *Climate change in the Philippines*. Retrieved from https://dilg.gov.ph/PDF_File/reports_resources/DILG-Resources-2012130-2ef223f591.pdf.

PAGASA (Philippine Atmospheric, Geophysical and Astronomical Services Administration). (2018). *Observed climate trends and projected climate change in the Philippines*. Retrieved from https://icsc.ngo/wp-content/uploads/2019/07/PAGASA_Observed_Climate_Trends_Projected_Climate_Change_PH_2018.pdf.

Pescaroli, G., & Alexander, D. (2015). A definition of cascading disasters and cascading effects: Going beyond the "toppling dominos" metaphor. *Planet@Risk, 3*(1), 58–67.

PSA (Philippine Statistics Authority). (2016). *Philippine population density (based on the 2015 census of population)*. Retrieved from https://psa.gov.ph/content/philippine-population-density-based-2015-census-population.

Rebs, T., Brandenburg, M., & Seuring, S. (2019). System dynamics modeling for sustainable supply chain management: A literature review and systems thinking approach. *Journal of Cleaner Production, 208*, 1265–1280. https://doi.org/10.1016/j.jclepro.2018.10.100

Ritchie-Dunham, J. L., & Méndez Galván, J. F. (1999). Evaluating epidemic intervention policies with systems thinking: A case study of dengue fever in Mexico. *System Dynamics Review, 15*(2), 119–138. https://doi.org/10.1002/(SICI)1099-1727(199922)15:2<119::AID-SDR163>3.0.CO;2-G

Saavedra, M. R., Fontes, C. H. D. O., & Freires, F. G. M. (2018). Sustainable and renewable energy supply chain: A system dynamics overview. *Renewable and Sustainable Energy Reviews, 82*, 247–259. https://doi.org/10.1016/j.rser.2017.09.033

Sayyadi, R., & Awasthi, A. (2020). An integrated approach based on system dynamics and ANP for evaluating sustainable transportation policies. *International Journal of Systems Science: Operations & Logistics, 7*(2), 182–191. https://doi.org/10.1080/23302674.2018.1554168

Sen, A. (1981). *Poverty and famines: An essay on entitlement and deprivation*. Oxford: Oxford University Press.

Sharpe, A., & Andrews, B. (2012). *An assessment of weighting methodologies for composite indicators: The case of the index of economic well-being*. Centre for the Study of Living Standards Research. Report No. 2012-10. Retrieved from http://www.csls.ca/reports/csls2012-10.pdf.

Solecki, W., Leichenko, R., & O'Brien, K. (2011). Climate change adaptation strategies and disaster risk reduction in cities: Connections, contentions, and synergies. *Current Opinion in Environmental Sustainability, 3*(3), 135–141. https://doi.org/10.1016/j.cosust.2011.03.001

Stave, K. A. (2003). A system dynamics model to facilitate public understanding of water management options in Las Vegas, Nevada. *Journal of Environmental Management, 67*(4), 303–313. https://doi.org/10.1016/S0301-4797(02)00205-0

Sy, C., Bernardo, E., Miguel, A., San Juan, J. L., Mayol, A. P., Ching, P. M., … Mutuc, J. E. (2020). Policy development for pandemic response using system dynamics: A case study on COVID-19. *Process Integration and Optimization for Sustainability, 4*(4), 497–501. https://doi.org/10.1007/s41660-020-00130-x

UNDRR (United Nations Office for Disaster Risk Reduction). (2021a). *Terminology: Disaster risk.* Retrieved from https://www.undrr.org/terminology/disaster-risk.

UNDRR (United Nations Office for Disaster Risk Reduction). (2021b). *Terminology: Hazard.* Retrieved from https://www.undrr.org/terminology/hazard.

UNISDR (United Nations International Strategy for Disaster Reduction). (2015). *Sendai Framework for Disaster Risk Reduction 2015–2030.* Retrieved from https://www.preventionweb.net/files/43291_sendaiframeworkfordrren.pdf.

Vafa-Arani, H., Jahani, S., Dashti, H., Heydari, J., & Moazen, S. (2014). A system dynamics modeling for urban air pollution: A case study of Tehran, Iran. *Transportation Research Part D: Transport and Environment, 31,* 21–36. https://doi.org/10.1016/j.trd.2014.05.016

Van Vuuren, D. P., Edmonds, J., Kainuma, M., Riahi, K., Thomson, A., Hibbard, K., … Rose, S. K. (2011). The representative concentration pathways: An overview. *Climatic Change, 109*(1), 5–31. https://doi.org/10.1007/s10584-011-0148-z

Wang, J., He, Z., & Weng, W. (2020). A review of the research into the relations between hazards in multi-hazard risk analysis. *Natural Hazards, 104*(3), 2003–2026. https://doi.org/10.1007/s11069-020-04259-3

5

Resilience planning in antagonistic communities

Stephen Buckman

Clemson University, Greenville, South Carolina, United States

1. An introduction to the planner's role in US communities

As planning and urban-studies graduate students in the United States, we had been taught that the community in which we should strive to live is one that is high density and adheres to more walkability, two of the major tenets that create a more sustainable urban environment. In turn, we leave our graduate programs holding our urban-planning Bible with its Old and New Testaments: the Old Testament being Jane Jacobs's *Life and death of great American cities* (1992) and the New Testament being Andres Duany and colleagues' *Suburban nation* (Duany, Plater-Zyberk, & Speck, 2000). With our planning Bible, we head out into the world to proselytize the ways of good sustainable planning and development. Yet, on the way to planning heaven on Earth, we inevitably run into a major problem that makes us question our own beliefs. We run into the issue of what to do when we enter a community that does not believe in the Gospel of density and sustainability and, in turn, simply does not want us there, to begin with.

Although we had been taught and, as educators and practitioners, we continue to teach planning orienting students that walkability and density are the ideals, many community members, especially in the United States, do not see it this way. Rather, much of the US population remains beholden to the automobile and the idea of the single-family house. Thus, invariably when we enter communities of this nature, we hit a brick

Investing in Disaster Risk Reduction for Resilience. https://doi.org/10.1016/B978-0-12-818639-8.00007-7

wall of backlash to our ideas. This backlash is often deeply rooted in political ideologies (Lewis, 2015) and the ideal of private property, in that we should not be telling them what they can and cannot do with their own land. The very notion of private property is one of the tenets of the American law, and the American psyche directly confronts the ideas of density and sustainability that we had been trained in. Thus, there is a psychological predisposition that sustainability measures threaten people's ability to profit from and do as they wish with their property. It then becomes important to communicate with such communities the fact that sustainability measures can equate to greater financial returns from the property market.

But trying to do planning in communities of this nature becomes a task that is not easy by any measure. This is especially true when the ideas of strong private-property rights and the perceived evils of sustainable planning are further stoked by the Tea Party (a far-right wing of the US Republican Party that calls for low taxes and the reduction of government spending) and anti-Agenda 21 activists. The opinions that these groups support are built around the very biased belief that planners (the "Left") are there to take their land. In their view, planners come into their communities and neighborhoods not to help but to harm by taking control of their private property (Foss, 2018; Trapenberg-Frick, 2013).

Those of us who do this type of work for a living would steadfastly argue that this is not the reason why we are in the communities. Rather, we contend that we are there to help communities create environments that are more livable and support sustainable efforts to combat issues of economic uncertainty and polarization. In a sense, it then can be avowed that sustainability and resilience make financial sense. As we look to bridge these gaps, we are still often perceived as invaders who want to hinder their ability to use their property as they wish. As we face this head-on, we must find ways to crossover and show that we are not the enemy and that what we propose are ideas to help strengthen a community's built form and social fabric now and in the future. This is particularly important as climate change modifies the entire structure of how our built form operates.

In this chapter, I will discuss my experience working with conservative neighborhoods in the US Great Lakes to show how to go about working with antagonistic communities. Specifically, I explain how scenario planning helped bridge the gap by shifting the focus to what was important to them. By no means should this chapter be looked at as a blueprint or a panacea. Rather it should be viewed as a way of thinking about how we as planners, urbanists, environmentalists, and sustainability-minded developers can go into communities and work in a manner that allows them to understand that sustainability and resilience are the best modes of action. In this chapter, I will first look at issues of culture and climate change, groupthink, Agenda 21, and power dynamics concerning planning. Then I will present a brief analysis of the work we did with highly antagonistic communities that did not want us there doing resilience planning. We were able to bring the community members together through the use of scenario planning for climate change by showing them the economic benefits of resilience. By highlighting the benefits related to sustainable development, we adhered to the principles of the Sendai Framework's Priority 3, on the importance of investing in disaster risk reduction. By reducing risks, we could invariably show the community the financial savings of planning now for later.

2. The importance of community interaction in the planning process

2.1 Understanding the basis of community antagonism

As planners, when doing our job, we must understand that communities in the United States are not always as we would like to believe but more and more they are becoming homogeneous. Historically, community homogeneity was predicated on economic and racial conditions that influenced where, how, and with whom people could live. Sometimes, these conditions were imposed by governmental policies at the federal and local levels, such as housing laws (especially redlining, which directed certain racial groups into particular home-buying areas), and immigrant quarters construed by ethnicity, race, and poverty. Yet, what is increasingly becoming prominent in today's homogenous communities is that they are not based on need but rather on choice. Sorting by choice has become a variable that we should look at when working with such communities. For instance, what kind of sorting? Why are they sorted? Habermas would contend that this is problematic in that sorting is a result of modern society, comprising competing traditions and cultural groups with different conceptions of the good. Hence, shared values are more likely the source of conflict in modern multicultural societies than they are key factors to their resolution (Finlayson, 2005; Habermas, 1985).

In his book *The big sort* (2009), Bill Bishop showed that place matters more than ever when talking about issues of trying to put forth different views of the world. This is because people are settling in places where they feel comfortable and they have the economic wherewithal to better organize themselves in areas that share the same political beliefs. The results are groupthink and self-imposed gerrymandering creating an echo chamber of ideas and propaganda. People are looking more and more for validation of their own beliefs, and sorting into communities further enables this to occur. In many communities, the voice of moderation and the middle is losing prominence with the dominant structural viewpoint becoming more extreme and present.

Whereas Bishop (2009) discussed why people are sorting into like-minded communities, Hoffman (2015) further highlighted the process of sorting in relation to preconceived views of climate change. What Hoffman (2015) showed in *How culture shapes the climate change debate* is that many people understand the issue of climate change but are intentionally avoiding the discussion. Hoffman (2015) also pinpointed that people inherently understand that climate change is happening but because of one's cultural identity, political convictions, and religious beliefs, they are careful to present or disregard scientific facts that do not fit their worldview. For instance, evangelical Christians, by their very nature, cannot believe that humankind has an impact on the climate as this would destroy their worldview that God is all-knowing and all-creating. In turn, for many of them, although they know that climate change is occurring, it is God's will, and hence there is no need to become more sustainable. Sorting into like-minded communities allows for the construction of fundamentally different moral institutions built around the perceived right way to live (Lewis, 2015). By doing this, people can further insulate their worldviews and cultural biases toward the climate-change debate.

The issues of sorting and the worldviews that predominate in these communities can also be a product of the ideas of bounded rationality. There is only so much humans can take in at once, entailing that we choose which information is most needed. The sorting that we see may as much be a survival mechanism in a hyperinformation world as a pure pushback of the other. What sorting of this nature further establishes is a type of activism driven by antidevelopment coalitions.

These coalitions, especially around sustainable development being a threat, are further stoked by the conservative Tea Party and anti-Agenda 21 activists, which Trapenberg-Frick envisioned as the new normal (Trapenberg-Frick, 2013). Both organizations perceive the state as the enemy and planners are, in their eyes, representatives of the state. Even though Agenda 21, signed in Rio de Janeiro in 1992, is a nonbinding UN action plan, anti-state organizations see it as an initiative to take control of the land and take away private property. They believe that it fosters higher urban densities leading to stack-and-pack housing. This equates in their minds to poverty, an attack on the middle class, and the breakdown of the very fiber of a good and just society. Trapenberg-Frick, Weinzimmer, and Waddell see the current opposition as being "embedded in a history of conservative concerns and growing conservative movement focused on the preservation of individual rights and the reduction of government" (2015, p. 212). Ironically, the very communities that are railing against governmental intervention are also heavily subsidized by the US government. This is particularly true for small rural communities throughout the country, such as Grand Haven Michigan.

Therefore, the place becomes important. In her 2016 book *The politics of resentment*, Katrine Cramer showed this importance as a lens through which people interpret their political beliefs, and, on the flip side, direct their attention to places where they feel others are eating more than their share of the economic pie. What permeates through Cramer's book is the idea of the fear of the other, and her subjects often see the other in terms of the "urban" who can easily be discerned to mean the minorities. The rural belief, which is at the core of her argument and holds in many rural towns, follows the ethos that "there is money in the economy, money to help people out. But not in their community" (Cramer, 2016, p. 178), as governments direct that money to urban locations. These communities are in general heavily opposed to government intervention or financial assistance as they see these not filtering down to them. Yet, the very perpetuation of these communities depends upon government subsidies and assistance. This viewpoint holds with many social theorists, such as Habermas, who argued that people vote against their self-interest because they hold false beliefs about what their true self-interests are and, unbeknownst to themselves, they behave irrationally. The result is that people are being funneled by economic and administrative systems into certain patterns of rationality (Finlayson, 2005; Habermas, 1985).

In this context, much of what is then occurring is a shift in power dynamics or a complete loss of power, or, more importantly, the feeling of a loss of power. For Habermas, communication and how we communicate are key deciders in relation to power dynamics. Habermas contended that language matters above all else as human actions are primarily coordinated by speech and language (Finlayson, 2005; Habermas, 1985). Power roles are mediated through the ways that such power dynamics are represented in language. As language involves the triangulation of the speaker, the hearer, and the world, any theory of language and power dynamics must take all three into account (Finlayson, 2005; Habermas, 1985).

Forester echoed this very notion in his seminal work *Planning in the face of power* (1989). A key tenet of his book is that planners must recognize that what gets done heavily depends on what gets said, how it is said, and by whom. Language, in Forester's mind, much like in Habermas's, is the driving force within planning. Through language and its dynamics, planners can thus understand how power relations shape the planning process and in turn improve the quality of their analyses, and empower citizens and community action (Forester, 1989). Along with language, planners need to listen and understand that how the language is being used matters. The issue for many planners is that they were not trained to listen. Rather they were trained in technocratic skills, but failing to listen denies a common membership in a shared world (Forester, 1989). As Forester (1989) further attested, to come to a mutual understanding between planners and the concerned community depends on the satisfaction of four main criteria: comprehensibility, sincerity, legitimacy, and accuracy. Balancing and embracing these principles through language help to bridge the political and antagonistic gap.

2.2 Overview of scenario planning

Understanding that language and communication are key aspects of successfully planning for resilience in antagonistic communities, it becomes important to find ways to create constructive narratives between opposing community members and planners. A strategy that urban planners working in such communities can employ to create a dialog and shift power dynamics is scenario planning, which allows for mutual options and decisions via debate and communication (Oteros-Rozas et al., 2015; Walton, O'Kane, & Ruwhiu, 2019).

Scenario planning can be used in the face of uncertainty to strengthen plans through various mechanisms and techniques. Each of these techniques plays into the way that planners can formulate scenarios to facilitate alternative futures and shift power dynamics. A description of a scenario is best situated within the historical background of "future" planning techniques, such as visioning and forecasting (Beach & Clark, 2015; Mirti & Hawken, 2020), which are often key components of a community's master plan or comprehensive plan.

Similar to these future-decision techniques, scenario planning looks to establish and build a narrative of the future. Invariably, scenarios are the archetypical product of future studies (Bishop, Hines, & Collins, 2007). The key distinction between scenario planning and the aforementioned planning techniques is that the former deals almost entirely with uncertainties. Scenario planning does not try to predict the future or forecast what is going to happen; rather the scenarios are projections for uncertain futures (Mohmoud et al., 2009; Smith, 2007).

The strength of scenario planning within the public-sector planning sphere is its ability to create multiple narrative stories of the future that aid decision-makers and communities to visualize development paths (Avin, 2007; Chakraborty, Kaza, Knaap, & Deal, 2011; Cummings, 2007; Hopkins & Zapata, 2007). Narrative stories around uncertain futures help communities to test policies and prioritize strategies for policy-making. They also demonstrate to stakeholders key future conditions (Holway et al., 2012), giving a community a way to process the future in the present (Harwood, 2007). The created stories are ideally rooted in verifiable economic, social, and environmental data about the present and expectations of the future (Cummings, 2007), by identifying how external conditions shape the community through multiple variables and perspectives (Harwood, 2007). Furthermore, scenario planning allows for communities to comprehend various uncertain outcomes to better adjust and adapt to the future. Unlike a forecast, which would lay out a constructed vision of the future for a community, scenario planning allows instead for multiple options to be discussed by community members and subsequently planned for by government officials. Hence, this allows for shifting power dynamics to the community away from a hierarchical government control.

A prerequisite of urban planning is weighing how changes under different economic, cultural, or environmental conditions would impact a community. As such, scenario planning can help this process by enabling planners and community members to understand how the combination of these future changes could affect their community. As scenario planning has become a more important tool with uncertainty becoming the norm, three subfields of urban planning that are traditionally concerned with uncertainty have used this technique extensively. First, transportation planning, which looks to uncertainties in travel demand; it has used scenario planning via community dialog to help encourage possible mobility futures ensuing from differing travel demands (Bartholomew, 2007; Bartholomew & Ewing, 2008; Zegras, Sussman, & Conkin, 2004). Second, environmental planning, which has led to better-informed decisions by bridging the gap between expert scientists, decision-makers, and communities (Peterson, Cumming, & Carpenter, 2003). Finally, climate-change planning, which allows planners to avoid the challenging sticky realm of forecasting and predicting; based on envisioned future conditions, communities can decide on their own about which is the best path to follow (Condon, Caverns, & Miller, 2009).

Another key point of scenario planning is for the communities to find what works best for them. In working with a community that is antagonistic and conservative, it is crucial to lead with what is important to them. Lewis (2015) showed that conservative-minded communities tend to value order and financial homogeneity over environmental considerations. Conversely, liberal-minded communities are more inclined to value denser communities as these equate to more diversity, and social and environmental equity. In turn, in many cases, for conservative communities, planners must help lead communities down the path of finances to get to the ecological. This dovetails with the Sendai Framework's Priority 3 of investing in disaster risk reduction for resilience.

Communities engaged in resilience planning are creating value through physical-infrastructure risk reduction, sustainable use and management of ecosystems, and disaster risk reduction concerning tourism areas. What the case study in this chapter will highlight is that the scenarios we led the groups through were very much dependent on showing the financial impacts of doing nothing and how doing nothing could destroy their main source of income: summer lake tourism.

3. Subject community: Grand Haven Michigan

The community of Grand Haven, located on Lake Michigan, participated in the scenario-planning development focused on in this chapter. It represents the main summer-tourism spot for Grand Rapids, Michigan's second largest city. The community was about to revise its master plan, a process that must be done every five years in the State of Michigan. Together with two other small Great Lakes communities, Grand Haven elected to have researchers from the University of Michigan and Michigan Tech University, as well as Land Information Access Association (a nonprofit planning think-tank), to work with them to design a resilience master plan. Part of the effort to revise their master plan involved extensive community meetings, which in turn aided in creating scenario plans designed to highlight perceived climate futures and policy outcomes.

Throughout the master-planning process, which took roughly nine months, the researchers conducted surveys and interviews with the stakeholders, and held community meetings. Before fully engaging community stakeholders, we sent them a survey to garner their thoughts on subjects such as climate change, resilience, and trust in government. We used this initial survey as a baseline to determine where to start from. As a research mechanism, we also conducted a postsurvey after the entire planning process to understand how their views changed. The second course of action consisted of in-depth interviews with government officials and key stakeholders in the community. Thirdly, we conducted multiple community-planning meetings. These meetings gave the community not only an overview of resilience and the planning process but also a stake in the planning outcomes. During the planning meetings, we engaged the community in what we called Community Action Teams, a three-step process that involved community dialog to take a broad question and narrow it down to a specific answer.

As researchers and planners, we hoped that, by working with the community members to understand what was important to them, it would make for a more robust, valid, and usable scenario plan. In turn, other communities of similar size and structure within the Michigan Great Lakes basin could use it to help formulate their own plans without outside, expert help.

3.1 Community roadblocks

While we were embraced with open arms by city government officials in the three communities we were working with and by some community members, there was a vocal amount of residents who did not want us there and were testy when it came to issues of climate change. An official in the first community we worked with told us that the community believed that climate change was happening and that we could talk about all the aspects of climate change and how its impacts were being felt. Yet, he emphasized that if we used the term "climate change" we would be dead in the water. Although they understood the basic ideas of climate change, the term itself was a trigger word that went against their very worldview and connoted a government takeover of their rights.

The inability to use the term "climate change" constituted a strong language roadblock that we had to overcome. We needed to find a way to get past this without destroying its basic understandings. Through internal deliberation, we came up with the terms "global weirding" and "climate weirding" interchangeably, which became the accepted lingua franca for our discussions throughout many of the open community meetings.

Another issue that we confronted from the outset was the vitriol within the actual community meetings. For instance, in one of our first meetings with one of the community representatives, the Tea Party and anti-Agenda 21 activists nearly came to blows with one of the city council's members. On top of that, our meetings were watched over by the author of an anti-Agenda 21 blog who would generally write a scathing review of what we were saying and proposing. These issues presented us with difficulties in finding a way to get the community members to talk with each other on a more civil and open plane. We looked to bridge these gaps via scenario planning, which at its heart is about opening veins of a narrative.

4. The scenario-building process and structure

Through discussions with community members from all three communities, local officials, and the research team, we compiled a grouping of climate futures and management options to populate the scenario chart. These variables included three climate futures, namely "Lucky," "Expected," and "Perfect storm" futures. They also comprised three management options: maintaining the present building and infrastructure policy, a build-out under the community's zoning code (build-out to 60% of allowable zoning expansion for this study), and a build-out under best management practices (avoidance regulations designed by the research team at the University of Michigan).

Under the "Lucky" climate future, the Great Lakes' water levels would continue to stay relatively low. On the other hand, under the "Expected" climate future, the water levels would continue to fluctuate according to long-term decadal patterns, including recent extreme storm events. There would also be more frequent large storm events than experienced in the past. Finally, under the "Perfect storm" climate future, the Great Lakes' water levels would continue to fluctuate according to long-term decadal patterns, consistent with the assumptions made by the US Federal Emergency Management Agency (FEMA) and the research team for the "expected" future. In addition to those assumptions, because of the foreseen increased frequency and intensity of storms, the shoreland flood-prone areas as delineated by FEMA's "500-year storm" event would become a part of the "100-year storm" event.

As regards the management assumptions, we first considered the maintenance of current practices imposed by each municipality. In this option, the community would manage land use in the same manner that it is currently doing, so without additional growth. This management option essentially provided the community with a snapshot of what it currently has in place. Alternatively, under a full build-out option, the community would undergo a full build-out allowable development according to its existing zoning code (at a 60% capacity to account for yards, driveways, and additional needed infrastructure, among others). New homes might be built in flood-prone areas. Finally, under best management practices (BMPs), the community would adopt and implement avoidance measures to preserve natural resources, protect private property, and manage the fiscal budget.

5. Achieved scenario plan

By combining the three climate futures and three management options, we designed a scenario plan for the community, which could be used by other similar communities as a blueprint. The achieved planning chart was composed of nine squares with the combinations of climate future and management option. Each combination could then be used to analyze key potential impacted outcomes in several domains. These include shoreline-management results (number of properties at risk and number of properties using shoreline hardening), critical facilities (number of facilities at risk of damage and specific facilities impacted), ecological outcomes (changes in the area and quality of wetlands and in impervious surfaces), and fiscal-impact analysis (the mean state-equalized value of residential, commercial, and government buildings).

The first step in populating each square was a quality set of base maps that visually showed the three future climate conditions—see Buckman, Arquero de Alarcón, and Maigret (2019) for further description. Whatever those climate conditions would be, the climatic base maps would be instrumental to understanding them, in that we built the rest of each scenario square. From these base maps, we superimposed further maps such as a zoning build-out or best-management wetland protection, forming a visualization of each scenario square. For example, Fig. 5.1 shows the three climate futures for the Grand Haven Michigan Community (Grand Haven City and Grand Haven Township).

Our study for the City of Grand Haven exemplifies how it was possible to get an understanding of what the impact of the three future climate conditions would be on the community. As Fig. 5.1 shows, flooding was naturally most likely to occur along the lakeshore and the river. By examining the map, the community could garner from the different flood scenarios what impacts those floods would have. The map shows expected impacts on properties at risk and how different management options could affect them, how many critical facilities would be a risk, the ecological outcomes, and lastly the overall fiscal impact from each flooding scenario.

5.1 Impacted land areas

By viewing the climate-impact map and transposing parcel and land-use maps on top of it, one can identify how much area and how many parcels and their types would be impacted by flooding. The results of climate futures on the City of Grand Haven show that flooding can menace from 136 to 245 ha, from 12% to 17% of residential parcels, and from 23% to 34% of the publicly owned parcels. Furthermore, the number of flood-impacted structures ranges greatly. For instance, if current policies hold, it is expected that between 57 and 256 structures would be impacted and a build-out could see between 207 and 497 structures, whereas implementing BMPs would decrease that to the 59–305 range.

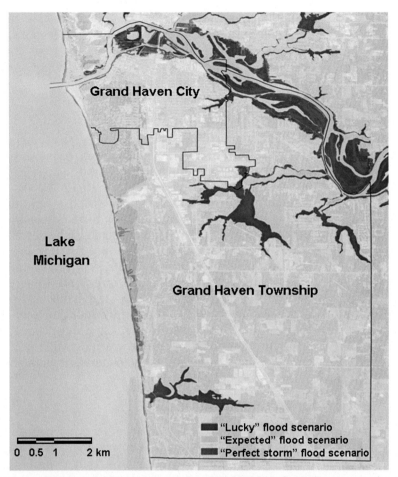

Figure 5.1 Grand Haven Michigan Community's inundation map, based on three climate scenarios. Based on Buckman, S., Arquero de Alarcón, M., & Maigret, J. (2019). Tracing shoreline flooding: Using visualization approaches to inform resilience planning for small Great Lakes communities. *Applied Geography, 113.* https://doi.org/10.1016/j.apgeog.2019.102097.

5.2 Fiscal-impact analysis

The fiscal analysis, which is often the portion of the scenario most anticipated and used by government officials, showed stark contrasts between future climate impacts and management options. The current structural option shows that the expected losses range from USD6 million to USD89 million. Conversely, if the community is built out, losses could range from USD32 million to USD180 million, whereas the BMP option would lessen this impact to between USD6 million and USD155 million. Being the most used portion of the scenario plan, the fiscal analysis is also the most contentious one. To offset this contention, we formulated a range of cost and revenue projections. For instance, the approximate direct building damage for the "Expected" climate future under current management options is in the range of USD18 million to USD85 million.

5.3 Ecological impacts

The last area of analysis focused on the ecological impacts for Grand Haven, as this area of inquiry has a direct impact on tourism activities (fishing, boating, and beaches). These impacts are generally the most underappreciated ones; yet, they can substantially influence other aspects that the community holds in higher regard, especially in terms of fiscal outcomes. From a fiscal perspective, the destruction of ecologically sensitive areas can impact future land development and tourism, for instance. The results show that there is very little change in the affected wetlands under different climate conditions. Impervious surfaces and critical dunes are especially impacted in the "Lucky" and "Expected" outcomes. The impervious surfaces would jump from 13.8 to 24.3 ha and the critical dunes from 22.7 to 80.1 ha. The critical dunes' portion of the ecological impacts is especially concerning, as these areas are key touristic points.

6. Community input

It was important to put together a scenario-planning chart that aggregated the climatic impacts on the area and the possible management options. But this was only one of the key steps in the process to make a more robust plan that would cross political boundaries. The other component was the qualitative capacity of dialog. On top of working through the scenario-planning chart, we also had the community arrive at an understanding and take ownership of certain variables of its resilience-building master plan. To do this, we instituted what we referred to as Community Action Teams (CATs), composed of community stakeholders.

CATs echoed the charrette process that many communities go through to understand what aspects of their area are most in need via social interaction. CATs met over three sessions to identify what each group thought were the pressing issues of their community. Unlike the charrette process, which is usually a mix of different citizens coming and going, CATs were composed of the same subgroups over the three sessions. Each CAT had a few (six to ten) persons who would tackle a specific issue such as transportation, environment, or economic development.

During the first session, the discussion was very broad as to what the main issues were; the second focused on some key issues. Finally, the last session was further narrowed down to only two or three pressing issues and ways to tackle them.

Thus, unlike the standard community workshop charrettes, CATs allowed for more specific answers to important questions facing the community. Participation in the CAT process offered stakeholders the chance to talk about their community's unique perspectives, distinct identity, and cultural attributes that were often overlooked in the common planning process. The CATs' results derived from allowing the community to specifically tackle the challenges facing it and further helped to inform decision-making. The CAT outputs became instrumental in aiding the community to decide what scenario plan worked best for it as decisions were made through open dialog. The CAT process helped to sow the seeds of the dialog that would be needed for the entire scenario planning to be a success.

7. Concluding comments

In the end, what this brief example shows is that scenario planning is far from a predictor of future events; rather it presents the possibilities and impacts of future events. In so doing, it puts forward strategies that communities could follow to manage conflicting issues. As the planning field increasingly shifts toward addressing uncertainty, scenario planning is positioned to be a key instrument in planners' toolkit. Furthermore, given that it offers a myriad of options, scenario planning can help to open up a dialog between communities and elected and nonelected government officials. In turn, by offering insights, not as an either–or situation, scenario planning provides an avenue for community and government actors to work together to come to a mutual decision as to what best suits their area.

Understanding that the community of Grand Haven is especially dependent on tourism for its economic livelihood, one could reasonably argue that the greater the storm and its destruction, the more tourists would be discouraged to come to the town. Thus, any storm would have to be looked at not only in its immediate consequences but also in its future ripple effects. This was the key factor that the community focused on when considering the scenario plan and the CAT results. As veins of dialog had been opened via CATs and scenario planning, the community was more equipped to cross political and cultural bridges and to break through walls put up by forces such as the Tea Party and anti-Agenda 21 groups.

Even though the planning process partially broke these barriers in this case, by no means is scenario planning a magic bullet that will insulate communities from uncertainties. But what it does do is to supply a community with a menu of options that this can pursue to lessen and adapt to future uncertainties and give voice to various concerns. As climate change and the uncertainties that come with it in all communities become more of the planning norm, scenario planning has the potential to help communities adapt to future climatic conditions in a collaborative manner.

Acknowledgments

This chapter and the Retrofitting, Rehabbing, and Recoupling Great Lakes Shoreline research project were funded by the Water Center at the University of Michigan and the Michigan Department of Environmental Quality Office of the Great Lakes Coastal Zone Management Program. I also thank Richard Norton, Zach Rable, Maria Arquero de Alarcón, and Jen Maigret for their collaboration.

References

Avin, U. (2007). Using scenarios to make urban plans. In L. Hopkins, & M. Zapata (Eds.), *Engaging the future: Forecasts, scenarios, plans, and projects* (pp. 103–134). Cambridge, MA: Lincoln Institute of Land Policy.

Bartholomew, K. (2007). Land use–transportation scenario planning: Promise and reality. *Transportation, 34*, 397–412. https://doi.org/10.1007/s11116-006-9108-2

Bartholomew, K., & Ewing, R. (2008). Land use–transportation scenarios and future vehicle travel and land consumption: A meta-analysis. *Journal of the American Planning Association, 75*(1), 13–27. https://doi.org/10.1080/01944360802508726

Beach, D., & Clark, D. (2015). Scenario planning during rapid ecological change: Lessons and perspectives from workshops with southwest Yukon wildlife managers. *Ecology and Society, 20*(1). https://doi.org/10.5751/ES-07379-200161

Bishop, B. (2009). *The big sort: Why the clustering of like-minded America is tearing us apart.* Boston, MA: Mariner Books.

Bishop, P., Hines, A., & Collins, T. (2007). The current state of scenario development: An overview of techniques. *Foresight, 9*(1), 5–25. https://doi.org/10.1108/14636680710727516

Buckman, S., Arquero de Alarcón, M., & Maigret, J. (2019). Tracing shoreline flooding: Using visualization approaches to inform resilience planning for small Great Lakes communities. *Applied Geography, 113*. https://doi.org/10.1016/j.apgeog.2019.102097

Chakraborty, A., Kaza, N., Knaap, G.-J., & Deal, B. (2011). Robust plans and contingent plans: Scenario planning for an uncertain world. *Journal of the American Planning Association, 77*(3), 251–266. https://doi.org/10.1080/01944363.2011.582394

Condon, P., Caverns, D., & Miller, N. (2009). *Urban planning tools for climate change mitigation.* Cambridge, MA: Lincoln Institute of Land Policy.

Cramer, K. (2016). *The politics of resentment: Rural consciousness in Wisconsin and the rise of Scott Walker.* Chicago, MI: University of Chicago Press.

Cummings, R. (2007). Engaging the public through narrative-based scenarios. In L. Hopkins, & M. Zapata (Eds.), *Engaging the future: Forecasts, scenarios, plans, and projects* (pp. 243–260). Cambridge, MA: Lincoln Institute of Land Policy.

Duany, A., Plater-Zyberk, E., & Speck, J. (2000). *Suburban nation: The rise of sprawl and the decline of the American dream.* New York, NY: North Point Press.

Finlayson, J. G. (2005). *Habermas: A very short introduction.* Oxford: Oxford University Press.

Forester, J. (1989). *Planning in the face of power.* Berkeley, CA: University of California Press.

Foss, A. (2018). Divergent responses to sustainability and climate change planning: The role of politics, cultural frames and public participation. *Urban Studies, 55*(2), 332–348. https://doi.org/10.1177/0042098016651554

Habermas, J. (1985). *The theory of communicative action: Volume 1. Reason and the rationalization of society.* Boston, MA: Beacon Press.

Harwood, S. A. (2007). Using scenarios to build planning capacity. In L. Hopkins, & M. Zapata (Eds.), *Engaging the future: Forecasts, scenarios, plans, and projects* (pp. 135–154). Cambridge, MA: Lincoln Institute of Land Policy.

Hoffman, A. (2015). *How culture shapes the climate change debate.* Stanford, CA: Stanford University Press.

Holway, J., Gabbe, C. J., Hebbert, F., Lally, J., Matthews, R., & Quay, R. (2012). *Opening access to scenario planning tools.* Cambridge, MA: Lincoln Institute of Land Policy.

Hopkins, L., & Zapata, M. (Eds.). (2007). *Engaging the future: Forecasts, scenarios, plans, and projects.* Cambridge, MA: Lincoln Institute of Land Policy.

Jacobs, J. (1992). *Life and death of great American cities.* New York, NY: Vintage Books.

Lewis, P. (2015). Moral institutions and smart growth: Why do Liberals and Conservatives view compact development so differently? *Journal of Urban Affairs, 37*(2), 87–108. https://doi.org/10.1111/juaf.12172

Mirti, A. V., & Hawken, S. (2020). Using scenario planning to enhance coastal resilience to climate change: Community futures in the estuarine landscapes of Brisbane water, Central Coast, Australia. *ISPRS Annals of the Photogrammetry, Remote Sensing and Spatial Information Sciences, VI-3/W1-2020,* 51–58. https://doi.org/10.5194/isprs-annals-VI-3-W1-2020-51-2020

Mohmoud, M., Liu, Y., Hartmann, H., Stewart, S., Wagener, T., Semmens, D., … Winter, L. (2009). A formal framework for scenario development in support of environmental decision-making. *Environmental Modeling & Software, 24*(7), 798–808. https://doi.org/10.1016/j.envsoft.2008.11.010

Oteros-Rozas, E., Martín-López, B., Daw, T., Bohensky, E., Butler, J., Hill, R., … Vilardy, S. (2015). Participatory scenario planning in place-based social-ecological research: Insights and experiences from 23 case studies. *Ecology and Society, 20*(4). https://doi.org/10.5751/ES-07985-200432

Peterson, G., Cumming, G., & Carpenter, S. (2003). Scenario planning: A tool for conservation in an uncertain world. *Conservation Biology, 17*(2), 358–366. https://doi.org/10.1046/j.1523-1739.2003.01491.x

Smith, E. (2007). Using a scenario approach: From business to regional futures. In L. Hopkins, & M. Zapata (Eds.), *Engaging the future: Forecasts, scenarios, plans, and projects* (pp. 79–101). Cambridge, MA: Lincoln Institute of Land Policy.

Trapenberg-Frick, K. (2013). The actions of discontent: Tea Party and property rights activists pushing back against regional planning. *Journal of the American Planning Association, 79*(3), 190–200. https://doi.org/10.1080/01944363.2013.885312

Trapenberg-Frick, K., Weinzimmer, D., & Waddell, P. (2015). The politics of sustainable development opposition: State legislative efforts to stop the United Nations' Agenda 21 in the United States. *Urban Studies, 52*(2), 209–232. https://doi.org/10.1177/0042098014528397

Walton, S., O'Kane, P., & Ruwhiu, D. (2019). Developing a theory of plausibility in scenario building: Designing plausible scenarios. *Futures, 111*, 42–56. https://doi.org/10.1016/j.futures.2019.03.002

Zegras, C., Sussman, J., & Conkin, C. (2004). Scenario planning for strategic regional transportation planning. *Journal of Urban Planning and Development, 130*(1), 2–13. https://doi.org/10.1061/(ASCE)0733-9488(2004)130:1(2)

6

Systems thinking toward climate resilience

Olalekan Adekola[1] and Jessica Lamond[2]
[1]Department of Geography, York St John University, York, United Kingdom;
[2]Department of Architecture and the Built Environment, University of the
West of England, Bristol, United Kingdom

1. Challenges for resilience building

Building resilience has become a sought-after means to combat the effects of anthropogenic climate change and related hazards (Kaur et al., 2017; Torabi, Dedekorkut-Howes, & Howes, 2018). Global development and humanitarian agencies, as well as the findings of international research, encourage policy-makers to contribute large amounts of resources toward strengthening resilience to climate hazards (Hemstock et al., 2018; Schipper, Thomalla, Vulturius, Davis, & Johnson, 2016). Resilience is a broad and contestable concept. Yet, we use here the term "building resilience" to describe participatory processes aimed at the adaptation and preparedness of communities to withstand and rapidly recover from shocks (such as floods and droughts), including the potential to build back better. This is consistent with the Sendai Framework and the literature that emphasizes the importance of participatory processes, coproduction of knowledge, and the participation of stakeholders and decision-makers (Borquez, Aldunce, & Adler, 2017). Building resilience is fundamental if we are to improve wellbeing and accelerate progress toward the Sustainable Development Goal 11, on the promotion of sustainable cities and communities (Rodriguez, Ürge-Vorsatz, & Barau, 2018).

Resilience-building initiatives typically require collective action to achieve their goals (Berkes & Ross, 2013), but such collaboration can be difficult to secure. These initiatives usually alter the distribution of risk and resources with implications for equity. Thus, they encounter management challenges that may

Investing in Disaster Risk Reduction for Resilience. https://doi.org/10.1016/B978-0-12-818639-8.00004-1

compromise their overall effectiveness (Aldunce, Beilin, Handmer, & Howden, 2016). Navigating this complexity requires engagements that not only bring policy-makers and technical and nontechnical experts together but also create a platform for discussions and guarantees that there will be appropriate attention to "dissensus." This should also enable a holistic examination of stakeholders' interests and explore how they perceive and value their environments and capture information relating to the distribution of benefits and costs across stakeholders in the resilience-building process.

2. Building resilience through stakeholder engagement

Sharifi et al. (2017) draw a distinction between the stages of resilience building, including design, implementation, and monitoring, all of which combine to improve the adaptive capacity of urban systems to climate hazards. The process of systematically determining the root causes of vulnerability, identifying solutions, and mapping them is also fundamental to developing urban systems' adaptive capacity to deal with climate hazards (Ribot, 2013). An awareness of how different stakeholder groups approach each of these elements is a prerequisite for developing effective urban resilience. Many studies have identified stakeholder engagement as an important strategy for enhancing communities' resilience to climate hazards (Adenle et al., 2017; Tompkins & Adger, 2004). For instance, Begg, Callsen, Kuhlicke, and Kelman (2018) provided some examples of how stakeholder engagement fosters effective governance of community action concerning building resilience to flooding in the United Kingdom and Germany. However, the literature on climate-resilience engagement is fragmented (Gardner, Dowd, Mason, & Ashworth, 2009). It only offers a fair understanding of what climate-resilience engagement is and the essential issues on which to engage stakeholders regarding building resilience and adaptation to climate hazards. Our approach is that stakeholder engagement entails empowering key groups—similar to the third level on Arnstein's (1969) "ladder of participation"—to play a vital role in ensuring that the goals of resilience building are met.

Tompkins and Adger (2004) and Aldunce et al. (2016), among others, highlighted the importance of the active involvement of all stakeholders in the process of building resilience. Different groups hold valuable knowledge based on their interests and experiences and so can help identify any gaps or inconsistencies in the process (López-Marrero & Tschakert, 2011). Investing in stakeholder engagement can be cost-effective, increase awareness, and enhance the capacity to identify and implement risk reduction strategies (Burnside-Lawry & Carvalho, 2016). This chapter explores the practice and the value of using a systems approach in stakeholder engagement to support resilience building.

As disaster risk reduction for resilience navigates a period of urgent importance, the lack of useful frameworks makes it difficult to yield a useful set of recommendations to guide specific actions such as climate-change engagement processes. As pointed out by Gardner, Dowd, Mason, and Ashworth (2009, p. 5), "climate change and climate adaptation have some features that make engagement on these topics particularly problematic." These include issues around misinformation and skepticism about climate change, reactions to uncertainty, variations in the capacity for long-term planning, and people's future objectives and goals. In many cases, the value of stakeholder engagement is suboptimized.

Although there exist specific frameworks aimed at supporting stakeholder engagement on climate adaptation (see, for instance, Conde, Lonsdale, Nyong, and Aguilar (2005), and Gardner et al. (2009)), these often do not provide a guide on the nature of the topics to be covered in the process. Addressing complex challenges requires systems thinking that, to date, has not been fully employed in guiding the discussions on stakeholder engagement for climate-change resilience building. Such discussions should go beyond the linear consideration of causes and solutions. They should also explore dynamic, multilevel, and often complex interactions that affect climate-resilience building and are of primary concern to stakeholders (Ford et al. 2013; Viner et al., 2020). For example, occurrences and decisions made in one part of the system (e.g., tackling drought at the national level) might impact other valued outcomes or even create new vulnerabilities (e.g., related to flooding at the local level). A holistic approach that recognizes the interconnectedness of the issues affecting development processes and the environment is fundamental to building resilience in communities (da Silva, Kernaghan, & Luque, 2012; Yokohata et al., 2019). The systems approach is a model that enables this exploration as it encompasses the whole picture, highlights the broader context, and considers interactions among multiple levels. Moreover, it acknowledges dynamic shifts that occur over time and encourages collaboration among stakeholders (Fast & Rinner, 2014).

Recent studies, such as the one carried out in 2012 by the Asian Cities Climate Change Resilience Network, have attempted to promote a systems approach to meeting the challenges of urban climate-change issues (da Silva et al., 2012). Yet, how authorities can extend this approach to the stakeholder-engagement process remains unclear. Despite the strong consensus on the need to involve all the stakeholders in the process of building resilience to climate change (Aldunce, Beilin, Howden, & Handmer, 2015), there is no established framework for doing so, particularly in developing countries. Few empirical studies have focused on how practitioners structure resilience ideas as they engage stakeholders in discussions aimed at building resilience to climate hazards. This gap effectively leaves practitioners without any scientifically sound idea of how to structure such engagement. This is not necessarily due to the organization of the engagement but rather to the nature of the discussions during the process.

This chapter uses the content of the discussions between stakeholders during climate-resilience workshops to help identify themes that can steer similar engagement approaches toward encompassing the entire resilience-building process. Understanding these holistic themes is particularly important as societies involve stakeholders in building resilience to ensure that we are getting the best out of these engagements. Therefore, the purpose of this chapter is to offer a practical approach that can structure stakeholders' discussions through "systems thinking" (Findeisen & Quade, 1997). This plugs a key gap in the stakeholder literature on the need for a systems-thinking approach that can capture complex and dynamic issues in stakeholder engagement. Our goal was to promote a holistic knowledge exchange that moves away from a reductionist mindset to a systemic one in the design and implementation of climate-resilience strategies.

3. Stakeholder engagement for resilience building as a complex system

Group discussion is an effective way to engage stakeholders in addressing decision-making problems (Gramberger, Zellmer, Kok, & Metzger, 2015). It allows stakeholders to exchange information and ideas and integrate experiential knowledge to enhance the overall outcome (Mondino, Scolobig, Borga, & Di Baldassarre, 2020). However, engaging multiple stakeholders with varying agendas, discourses, preferences, work routines, norms, values, interests, and power in a process of building resilience can be complex (Aldunce et al., 2015; Lechner, Jacometti, McBean, & Mitchison, 2016). The management of climate-change-induced risks can result in a conflict situation from resource shortages and antagonistic feelings. The process also does not sit comfortably in traditional technical-expert-led policy-response regimes. An aspect of this complexity is the multi-faceted (from drought to flood) and multidimensional (from local to global) nature of climate risks, which has interlinked short-, medium-, and long-term aspects and sometimes unknown outcomes. Thus, engaging stakeholders in managing climate risks is an intricate and dynamic endeavor, given the potential uncertainty and ambiguity surrounding how each stakeholder views and reacts to an issue.

Systems thinking deals with understanding these interactions and interconnections, and identifying the leverage points to solve societal problems (Williams, Kennedy, Philipp, & Whiteman, 2017). Our idea of systems approach goes beyond looking at individual stakeholders and, instead, focuses on the contents and patterns of interactions between all the stakeholders as a part of the whole process. This approach contrasts with the traditional reductionist thinking that studies systems by breaking them into separate smaller elements for ease of analysis using simple relationships. The potential shortcoming of reductionist thinking is to have practitioners work in silos without keeping the larger societal context in view. Therefore, it has limited effectiveness in addressing the ever-increasing complexity of problems (including climate-related hazards) being faced by many communities.

To better understand and tackle complex and persistent problems, McCabe and Halog (2018) promoted systems-thinking techniques as a potential entry point for stakeholders' inclusion and knowledge coproduction. There is a growing body of resilience-building literature that draws on concepts from systems thinking (Cavallo & Ireland, 2014; Rasul & Sharma, 2016). By viewing climate-resilience building as a system, we can reach an integrated understanding of all interests and see how interlinked the impacts are, rather than dividing our understanding into separate areas of expertise. Systems thinking thus provides practitioners with an approach to map and understand interlinkages and their dynamics.

4. Exploring systems thinking for climate adaptation in Nigeria

This chapter is a reflective writing based on insights from research experiences during the project "Adaptation to enhance climate resilience of urban infrastructure in Nigeria," part of the Urbanisation Research Nigeria program funded by the UK Department for International Development. As important concentrations of economic assets, populations, and infrastructures, cities and urban areas are particularly vulnerable to climate-related hazards. Therefore, climate hazards pose serious threats to urban infrastructures, quality of life, and entire urban systems (Hunt & Watkiss, 2011).

After seeking and receiving full ethics approval for this study from the lead institution, we collected data using stakeholder workshops and focus-group discussions conducted in two Nigerian cities. We preselected four cities (Calabar, Enugu, Lokoja, and Makurdi) with differing climare-risk profiles and urban-planning histories. We then carried out a detailed literature review leading to the selection of Calabar and Makurdi for the data collection. These cities represent two ecological zones, one from each of the two broad belts of vegetation types in Nigeria (forest and savannah). Both are susceptible to coastal and fluvial flooding as well as water shortage, windstorms, and urban heat.

The aim of the workshops, held in June 2016, was to understand the concept of adapting infrastructure for resilience in urban Nigeria. We defined infrastructure broadly, as a set of systems, such as transport and communications, which enable the functioning of households and businesses. The workshops provided an opportunity to discuss such issues with a wide range of stakeholders, including academics, government, the media, community leaders, security forces, and NGOs. Based on insights obtained from the literature review and the scoping studies, we identified and mapped relevant city-wide stakeholders into five categories, namely federal-government actors, state-government actors, local-government actors, CBOs/CSOs/NGOs and multilateral organizations, and the formal private sector. Through this initial process and based on our knowledge, we then chose specific organizations to participate. We invited relevant individuals from each organization to ensure as much representativeness of the climate-hazard-management community as possible. Those invited were involved in diverse types of decision-making, operational, and knowledge roles, and we ensured that the purposively selected participants represented different groups.

The first one-day workshop took place in Makurdi with 58 participants and the second one took place in Calabar with 62 participants. A neutral venue (a hotel) and the direct invitation from the research team identified the workshops as open to all relevant stakeholders. The workshops consisted of presentations by invited speakers selected from relevant stakeholder organizations, followed by open plenary discussions. Smaller-group sessions facilitated by the project team were subsequently held and each group nominated members to present a summary of these discussions to the wider workshop audience. This setting was to encourage discussions and questions about key issues without any limits. The topics pertained to institutional arrangements for climate-hazard adaptation (especially flooding), as well as the challenges and expectations for their implementation. The open nature of the discussions allowed participants to build consensus but also express divergent views.

Alongside the workshops, we conducted two focus-group discussions (one in each city) in October 2016, to interface with community members and business leaders. We took detailed notes of the workshops and focus-group discussions, and then inductively analyzed the transcripts, through an open coding using traditional text analysis (Bright & O'Connor, 2007) to identify the themes that the participants considered essential for building resilience. We coded each theme according to the terminology used in existing systems-approach literature (Findeisen & Quade, 1997).

5. Capturing stakeholders' views of climate hazards and resilience building

We found that all the stakeholder groups were eager to discuss and work to build resilience to climate-related hazards mainly because of the direct impacts on human life, health, and livelihoods. When we asked about what concerned them most among the impacts of these hazards, direct economic losses (such as loss of income), and the degradation of water quality featured most prominently. The other impacts in order of concern were damage to property, ecological damage, death, disrupted transportation, injury, and the separation of people. Some participants also identified the outbreak of epidemics (such as cholera) and hazards' psychological effects as other issues of concern. In particular, a major worry in Makurdi was the situation whereby displaced people are not able to return to their homes after a flood. In

Calabar, displaced people housed in schools, affecting their functioning and abruptly truncating school hours, were a main cause of apprehension.

That said, the stakeholders often held some differing perspectives on key aspects associated with building resilience. For example, households and businesses across both cities were unanimous in rating electricity and drainage as the most important infrastructures to be resilient. Conversely, community leaders rated more highly local infrastructures such as roads, schools, and health facilities, as well as water, but agreed that sewage and telecommunication systems were less important. Likewise, regarding adaptation preferences, government stakeholders felt top-down directional approaches were the most appropriate. Households and businesses also preferred the government to take direct action, through either state-led programs (Makurdi) or engineered solutions (Calabar). Community leaders more or less agreed with this ranking, whereas they also put drainage improvement and rechanneling watercourses as number one, but the enforcement of laws ranked last in Makurdi. These differences transcend mere problem identification and championed solutions.

6. The seven-step systems approach

Based on our discussions with the stakeholders and the issues raised, we identified and synthesized important steps that should guide stakeholder engagement when building resilience to climate hazards. Our synthesis builds on Findeisen and Quade's (1997) six steps of environmental-systems analysis for integrated environmental management, which emphasize the need for an iterative and holistic approach to achieving sustainable outcomes.

Overall, the key themes discussed can be collated into seven overarching steps: (1) the need to broaden the knowledge source, (2) problem identification, (3) system definition, (4) system synthesis, (5) system analysis, (6) identifying plausible futures, and (7) communication (Fig. 6.1). These steps have the potential for making resilience building more sustainable and ensure investment in efficient outcomes through holistic decision-making. These seven steps also illustrate an approach for gaining crucial and useful knowledge in planning resilience. In Table 6.1, we present the elements of our model showing previous studies that already highlighted them.

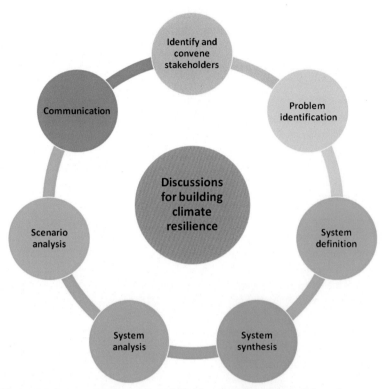

Figure 6.1 Seven steps to guide stakeholder discussions for building climate resilience. The authors.

6.1 Identify and assemble stakeholders

The workshop participants emphasized the importance of identifying and engaging in the discussions all the actors representing each stakeholder group. This included convening practitioners (such as planners, engineers, and architects), a broad variety of members of the local community, policy-makers, and decision-makers from particular sectors, geographic regions, and disciplines. Moreover, the workshop participants recognized that cross-sectoral collaboration is critical for building climate resilience due to the multifaceted nature of the challenges. They were aware that, at best, each of them focused on a single sector and such multisector collaborations are essential to gain a better understanding and enhance outcomes. Others viewed this as an avenue to facilitate knowledge sharing and collaboration at an organizational level: mobilizing community members and the private sector around a local vision for a resilient future.

Table 6.1 Summary of the steps of a systems-thinking discussion.

Stages	Comments	Literature
Convene stakeholders	All-stakeholder participation is critical from the onset for the outcome to be holistic, become socially relevant, and ensure that challenges such as meaningful communication and power structures are addressed. Discussions should help identify relevant stakeholders and their significance to the process of building resilience not just as a precondition to make the process participatory but also to ensure the full scale of participants in the system is engaged.	Aldunce et al. (2016), Hung, Yang, Chien, & Liu (2016), Tompkins & Adger (2004)
Problem identification	Identifying which problems are crucial to all stakeholders provides the clarity and transparency necessary to arriving at viable solutions and to ensuring that the process of building resilience is fair. Participants should be encouraged to indicate what they perceive the key problems to be, including those possibly external to the system. Such discussion provides a deeper understanding of on-the-ground realities that might inform a framework for considering the actions needed to build resilience.	Kernaghan & da Silva (2014), UNDP (2010)
System definition	With an understanding of the problem, participants should give serious consideration to how the things are linked. While climate hazard and resilience building may be the main topics for discussion, other external issues may be widespread and need to be considered. Given this link, discussions can help to design policies that tackle main and external situations simultaneously.	Fiksel (2003), Romero-Lankao, Gnatz, Wilhelmi, & Hayden (2016)
System synthesis	Discussions should enable participants to identify and be specific about solutions that are viable and sustainable within a local context. During such discussions, participants should be able to identify and debate possible unanticipated/unintended outcomes of the proposed solutions.	Saunders & Becker (2015), Tyler & Moench (2012)
System analysis	Discussions should go beyond giving participants time to suggest solutions; they should be encouraged to evaluate the suggestions especially on their novelty and usefulness to the local context, based on lessons from elsewhere.	Adger, Adams, Evans, O'Niell, & Quin (2013), Brisley et al. (2016)
Scenario analysis	Discussions should encourage participants to explore future conditions that cannot be readily predicted, envision potential futures, and explore alternative pathways to the	Bizikova, Rothman, Boardley, Mead, & Kuriakos (2015),

Table 6.1 Summary of the steps of a systems-thinking discussion.—*continued*

Stages	Comments	Literature
	desired ends. Such discussions can help highlight the trade-offs and synergies between present and future generations or among stakeholders that might be less evident when a more limited (static) suite of options is considered.	Gardner et al. (2009), Star et al. (2016)
Communication	Projects often "fail" because they do not clearly articulate and communicate their key aspects in an appropriate manner to the stakeholders. To enhance resilience building on a topic that is difficult for some groups to understand requires communications that enable all stakeholder types to visualize the end result, so as to work toward a common goal.	Aldunce et al. (2016), Jha, Brecht, & Stanton-Geddes (2015)

6.2 Problem identification

First and foremost, the stakeholders commented on the importance of identifying the most significant climate-related hazards confronting their cities. Our study revealed that the climate hazard of the greatest concern differed between the two cities and among the stakeholders within each city. For example, in Calabar, flooding was overall identified as being the climate hazard of greatest concern; in Makurdi, it was urban heat. However, in Calabar, while households and government stakeholders ranked urban heat as the second concern, business owners and NGOs ranked urban heat as being of equal concern to flooding.

The process of problem identification involved the participants pointing out the natural, social, economic, and political factors driving climate-related hazards. When asked about the causes of these hazards, the participants recognized a combination of both natural and human-made factors for flooding and urban heat. One of the major human factors identified as causing flooding is the cultural shift toward increased consumption of sachet water, which generates plastic wastes that clog drainage channels and prevent the free flow of water. For windstorm, however, the participants generally believed that the main causes were natural factors. This implies that the perception of the cause of a climate hazard can be highly issue-specific. The participants suggested that when the problems are not properly identified, seemingly innovative solutions could face bottlenecks in their implementation and erode trust among the stakeholders.

6.3 System definition

The workshop participants highlighted the importance of having a significant understanding of the interlinkages between the different sectors when discussing climate hazards. What was also clear from the discussions was the emphasis that the stakeholders put on the linkages between different urban infrastructures. For example, a participant pointed out that "*without electricity, every other thing is useless,*" indicating that much of the other infrastructure relies on electricity to function properly. This suggests that the stakeholders were not only identifying the problems but also defining the boundaries of the issues that they consider to hold equal importance. Such an appropriate system definition is essential for the identification of relevant solutions to the problem studied, as well as for determining the resource requirements (Findeisen & Quade, 1997). The discussions that fit participants' attempts at defining the system emphasized the spatial and temporal dimensions of the factors driving the problems, as well as the relationship between the problems and other challenges facing their society. For example, stakeholders in Makurdi emphasized the linkages between flooding and school closure and disease outbreaks. An appropriate system definition is essential for a holistic view of the issues at hand that allows for the identification of the most suitable solutions.

6.4 System synthesis

In addition to defining the problem and the boundaries of the issues at stake, other key themes emerging from the discussions were the identification of possible solutions for the problems and the exploration of the consequences of applying individual solutions and their combinations. The participants' suggested solutions were aligned with the defined system boundaries. For example, related to the institutional dimension, the workshop participants proposed the establishment of a new governmental agency mandated to coordinate land management. The participants were quick to also point out that such an agency should not be bureaucratic; rather it should ensure synergy and promote exchange and interaction between the stakeholders. Another suggestion was related to the provision to communities of hazard and resilience information by private-sector third parties, considered the stakeholders best positioned for this role. Since the participants identified diverse resilience-building options, most of the discussions also explored the benefits and trade-offs of individual and systemic solutions. The synthesis of the system in engaging stakeholders should promote the identification of holistic options that are available for building resilience, as well as their potentials and associated costs.

6.5 System analysis

The participants' discussions did not stop at identifying options that would aid the promotion of resilience building; they also drew comparisons with what has been applicable elsewhere, within Nigeria and internationally. For example, one of the financing mechanisms suggested in Makurdi was to raise internally generated revenue, following the approach adopted in Lagos (the largest city in Nigeria). Some of the discussions centered on why this has worked in Lagos and then on how this could be adapted to other local contexts. Similar debates occurred in Calabar. The participants noted the importance of creating an exchange network to share lessons learned that could be replicated elsewhere.

6.6 Scenario analysis

The participants were also forward looking, recognizing the uncertainty of future climate and development patterns, which required something resembling a scenario analysis. The majority opinion was that engaging stakeholders in other technical aspects of building resilience—such as exploring, qualifying, and reporting possible future changes—was also valuable. The participants viewed possible future developments in the management of climate hazards, within their cities and nationally, as important drivers but also outcomes of pursuing scenario analysis. They acknowledged that such a framework of discussions aimed to highlight possible consequences of the alternatives considered for some situations, that is, what would happen because of the actions suggested by other participants or what would happen without these actions.

6.7 Communication

The final key theme identified in the discussions was the need to communicate appropriately with all the stakeholders. The participants emphasized that communication and exchange of knowledge between stakeholders are currently poor. Some contended that most of the challenges militating against effective management of climate hazards have roots in *"the lack of communication."* The results also showed that communication is far from optimal and knowledge tends to be utilized ineffectively. For example, in Makurdi, workshop participants pointed out that the state emergency agency did not have ongoing contacts with the National Space Research and Development Agency, the Nigerian body that holds satellite-based data useful for decision-makers in disaster preparedness and response. The participants suggested that it is important to seriously consider to whom and in what ways the communication is done.

7. Putting the proposed systems approach into perspective

Based on our Nigerian experience and taking a systems-thinking approach, we proposed a set of steps to guide climate-resilience stakeholder-engagement processes. In these key steps, the inputs of various stakeholders are essential to building a holistic understanding of climate risks so as to reduce exposure and improve the adaptive capacity of urban systems. This includes bringing to the fore relevant issues that might not have been covered by the projects' objectives. Adopting this approach in our study generated an added understanding of the cultural (e.g., the problem of water sachets) and structural (e.g., the inefficiency in the communication) components of the urban system. These are vital prerequisites for identifying, designing, and evaluating the interventions to build resilience (Cristiano, Zucaro, Liu, Ulgiati, & Gonella, 2020). This way, all stakeholders can strategically and systematically develop solutions, ensuring that the effective ones are not left to chance.

In line with the systems-approach literature (Findeisen & Quade, 1997) and recent thinking in international development (UNDP, 2020), our findings suggest that the first step of the framework—identify and assemble stakeholders—is critical. Although having this seven-step structured framework to guide the resilience-building discussions, the most important requirement is to engage all the stakeholders from the onset. The ability to engage all the stakeholders to help identify and frame the key issues will be most decisive for the success of the resilience-building process. Such engagement should be active rather than passive and carefully consider who is involved and how, as these are crucial aspects for developing resilience in practice (Aldunce et al., 2016). Process conveners need to create spaces that ensure that all stakeholders are recognized and empowered. This calls for new approaches, such as the Learning Action Alliance framework (O'Donnell, Lamond, & Thorne, 2018), which can foster mutual learning rather than one-way communication.

The clarity of thought and direction created by the problem definition will go a long way to set the focus for the process. This can be useful to practitioners and stakeholders in cocreating initiatives that lead to effective results in a resilience-building project (Frantzeskaki, Hölscher, Wittmayer, Avelino, & Bach, 2018; Fraser, Dougill, Mabee, Reed, & McAlpine, 2006). Besides, if the problem is not clearly defined at the start, the goals can become fuzzy and superficial, and may lead to arguments later when considering the intended outcomes and strategies. Resilience-building practitioners should be aware that some stakeholders may try to define the problem from their points of view. And if one group succeeds in having the matter defined to suit their interests, it then becomes harder to resolve it from another perspective. The problem-definition step is, therefore, crucial to ensure that there are both unity of purpose and power balance among the stakeholders, through collective agency, which is central to the process. Also, a clear understanding of the problem at hand is important to ensure that it is correctly formulated, as analysis failures often derive from the attempts to solving the wrong problem. Problem identification provides a platform for stakeholders to start investigating a broad range of interventions and generating options.

The stakeholder discussions that define the systems' boundaries regarding all interlinked issues are also essential, as they can ultimately help designing policies that tackle main and adjacent situations simultaneously. As highlighted by Moser & Ekstrom (2010), the stakeholder discussions should recognize the importance of understanding the interlinkages between climate hazards and other key developmental issues, such as urban poverty. An essence of the discussions on system synthesis is the acknowledgment that adaptation can mean different things to different stakeholders (Bours, McGinn, & Pringle, 2014). As such, the discussions can enable the stakeholders to collaboratively identify—and thus avoid—possible unanticipated and unintended outcomes of the proposed solutions (Magnan, 2014).

Engaging stakeholders through discussions that invite them to evaluate the usefulness of resilience-building initiatives may help broaden their thoughts, promoting wider evaluations of these initiatives, rather than narrow views based only on the immediate local context. Striking an appropriate balance between the desire for context-specific solutions and an evaluation based on other settings is essential to create efficiencies while remaining effective in different urban sectors (Landauer, Juhola, & Klein, 2019). Cutter et al. (2008) equally emphasized the importance of understanding not just spatial but also temporal dynamics that may influence resilience building. This underscores the importance of discussions that encourage the stakeholders to envision and explore future conditions. Such discussions can help highlight the trade-offs and synergies between present and future generations or among stakeholders that might be less evident when a more limited suite of static scenarios is considered. Resilience building typically seeks solutions to wicked problems, in which multiple stakeholders with divergent values engage within complex and dynamic systems to achieve the common good. Effective communication in every step ensuring that all stakeholders are carried along and their views and perceptions are incorporated is also essential. The practitioners engaging stakeholders in building resilience should therefore adopt a communication strategy that enables all the actors to visualize the results.

Our study highlighted the differences in the stakeholders' viewpoints at each of the stages. Therefore, missing out any of the steps would mean that the stakeholders' inputs and consensus building could fall apart. Such differences reflect the abrogation of responsibility by groups (Bigg, 2013). For example, it can be surmised that people calling for more laws, regulations, and education are hoping that these will make others act more "responsibly" (Coleman, 2012) and avoid the need for them to act. Without a systems approach that identifies and recognizes such diversity, the resilience-building process could lead to the entrenchment of a blame culture that can escalate climate-hazard incidents toward catastrophic impacts.

Another important value of this seven-step framework is that, by ensuring that the knowledge of technical and nontechnical stakeholders is integrated across spatial and temporal scales, it recognizes and can reduce power imbalances in the resilience-building process by identifying and openly and collectively negotiating conflicts in perspectives. Despite its related implicit stance, the Sendai Framework is not vocal about power imbalances among stakeholders. This is why Dewulf et al. (2019) called for a power-sensitive resilience framework. Our framework emphasizes that accepted interests, viewpoints, and knowledge cannot belong to some of the stakeholders acting as superior to others. Otherwise, the ensuing process of power imbalance favoring some stakeholders at the expense of others could result in conflicts and prove expensive to manage.

Although the proposed framework is not the only path to addressing stakeholder participation in climate-risk resilience building, it does provide a useful checklist to assist practitioners in convening discussions and achieving consensus among stakeholders along the principal stages that may be missed during their engagement. Testing the robustness of this framework requires empirical studies to examine how those key stages might differ and under what circumstances. Furthermore, collecting empirical information while applying this framework would enable its evaluation and refinement. Overall, our study demonstrates the value of holistic frameworks and shows that climate-management stakeholders recognize the importance of such a comprehensive approach in resilience building.

8. Investing in systems thinking toward climate-change resilience

Engaging stakeholders in the process of building resilience to climate hazards presents an interesting example of conflict between multiple parties with differing concerns. Due to the very nature of the lengthy time horizons of resilience-building projects, the stakeholders of those projects may even change their positions and preferences over time. Such a complex and dynamic nature of stakeholder engagement requires a holistic approach with appropriate frameworks.

We presented the results of an examination of the key themes discussed during two stakeholder workshops aimed at building resilience to climate hazards in Nigeria. We then explored how the notion of systems thinking could be useful for practitioners engaging multiple stakeholders in such processes. Through its seven steps to systems thinking, this chapter shows how we can better structure and capture the interests and views of diverse stakeholders in understanding, designing, and communicating as regards climate adaptation. The identified key frames of stakeholder discussions move beyond the usual debate around cause, effect, and solution, to encompass exploring future scenarios and communication (see Chapter 5, by Buckman, in this volume). Overall, we propose that systems thinking constitutes a useful framework for helping policy-makers and practitioners to engage multiple stakeholders in building resilience.

The application of systems thinking to structure engagement around building resilience provides an opportunity to better understand complexity by viewing climate risk and the consequences of change as part of the overall system. This can enable practitioners to better contextualize their risk landscape and reduce power imbalances. This approach can also assist practitioners in identifying interlinkages between hazards and impacts. Discussing potential impacts leading to possible future scenarios can help to build adaptation approaches that meet the needs of the future. Another strength of this framework is that different environmental-systems analysis and tools can be incorporated into the process to provide some scientific validity and outlook to it. Likewise, iteration loops between the stages may be useful for further engagements, if the discussions at any stage give rise to changes in the system. The systems-thinking approach connects problem definition with communication more effectively.

The systems-thinking framework on its own is not a panacea, and its application may not mean that the challenges of engaging in complex resilience building will be straightforwardly resolved. Yet, it can enhance the development of a common understanding and help practitioners to structure the discussions and look out for key information. Finally, we recognize that the viability of using this framework must be demonstrated through examples from resilience-planning activities undertaken across different contexts. Furthermore, there is a need to consolidate the state of scholarly research on stakeholder engagement in climate-change adaptation. This should include studies that provide further insights into how to keep people actively engaged throughout the long process of resilience building.

References

Adenle, A. A., Ford, J. D., Morton, J., Twomlow, S., Alverson, K., Cattaneo, A., ... Ebinger, J. O. (2017). Managing climate change risks in Africa—A global perspective. *Ecological Economics, 141*, 190–201. https://doi.org/10.1016/j.ecolecon.2017.06.004

Adger, N., Adams, H., Evans, L., O'Niell, S., & Quin, T. (2013). *Human resilience to climate change and disasters: Response from University of Exeter.* Retrieved from https://royalsociety.org/ ~ /media/policy/projects/resilience-climate-change/parts1-20.pdf.

Aldunce, P., Beilin, R., Handmer, J., & Howden, M. (2016). Stakeholder participation in building resilience to disasters in a changing climate. *Environmental Hazards, 15*(1), 58–73. https://doi.org/10.1080/17477891.2015.1134427

Aldunce, P., Beilin, R., Howden, M., & Handmer, J. (2015). Resilience for disaster risk management in a changing climate: Practitioners' frames and practices. *Global Environmental Change, 30*, 1–11. https://doi.org/10.1016/j.gloenvcha.2014.10.010

Arnstein, S. R. (1969). A ladder of citizen participation. *Journal of the American Institute of Planning, 35*, 216–224. https://doi.org/10.1080/01944366908977225

Begg, C., Callsen, I., Kuhlicke, C., & Kelman, I. (2018). The role of local stakeholder participation in flood defence decisions in the United Kingdom and Germany. *Journal of Flood Risk Management, 11*(2), 180–190. https://doi.org/10.1111/jfr3.12305

Berkes, F., & Ross, H. (2013). Community resilience: Toward an integrated approach. *Society & Natural Resources, 26*(1), 5–20. https://doi.org/10.1080/08941920.2012.736605

Bigg, T. (Ed.). (2013). *Survival for a small planet: The sustainable development agenda*. London: Earthscan.

Bizikova, L., Rothman, D. S., Boardley, S., Mead, S., & Kuriakos, A. T. (2015). *Participatory scenario development and future visioning in adaptation planning: Lessons from experience. Part I*. IISD Working Paper. Retrieved from https://www.iisd.org/system/files/publications/participatory-scenario-development-future-visioning-adaptation-lessons-part-i.pdf.

Borquez, R., Aldunce, P., & Adler, C. (2017). Resilience to climate change: From theory to practice through co-production of knowledge in Chile. *Sustainability Science, 12*(1), 163–176. https://doi.org/10.1007/s11625-016-0400-6

Bours, D., McGinn, C., & Pringle, P. (2014). *Twelve reasons why climate change adaptation M&E is challenging*. Retrieved from https://www.ukcip.org.uk/wp-content/PDFs/MandE-Guidance-Note1.pdf.

Bright, M. A., & O'Connor, D. (2007). Qualitative data analysis: Comparison between traditional and computerized text analysis. *Osprey Journal of Ideas and Inquiry, 21*. Retrieved from http://digitalcommons.unf.edu/ojii_volumes/21.

Brisley, R., Wylde, R., Lamb, R., Cooper, J., Sayers, P., & Hall, J. (2016). Techniques for valuing adaptive capacity in flood risk management. *Proceedings of the Institution of Civil Engineers—Water Management, 169*(2). https://doi.org/10.1680/jwama.14.00070

Burnside-Lawry, J., & Carvalho, L. (2016). A stakeholder approach to building community resilience: Awareness to implementation. *International Journal of Disaster Resilience in the Built Environment, 7*(1), 4–25. https://doi.org/10.1108/IJDRBE-07-2013-0028

Cavallo, A., & Ireland, V. (2014). Preparing for complex interdependent risks: A system of systems approach to building disaster resilience. *International Journal of Disaster Risk Reduction, 9*, 181–193. https://doi.org/10.1016/j.ijdrr.2014.05.001

Coleman, J. (2012). *Take ownership of your actions by taking responsibility*. Retrieved from https://hbr.org/2012/08/take-ownership-of-your-actions.

Conde, C., Lonsdale, K., Nyong, A., & Aguilar, I. (2005). Engaging stakeholders in the adaptation process. In B. Lim, E. Spanger-Siegfried, I. Burton, E. Malone, & S. Huq (Eds.), *Adaptation policy frameworks for climate change: Developing strategies, policies and measures* (pp. 47–66). Cambridge: Cambridge University Press.

Cristiano, S., Zucaro, A., Liu, G., Ulgiati, S., & Gonella, F. (2020). On the systemic features of urban systems: A look at material flows and cultural dimensions to address post-growth resilience and sustainability. *Frontiers in Sustainable Cities, 2*(12). https://doi.org/10.3389/frsc.2020.00012

Cutter, S. L., Barnes, L., Berry, M., Burton, C., Evans, E., Tate, E., & Webb, J. (2008). A place-based model for understanding community resilience to natural disasters. *Global Environmental Change, 18*(4), 598–606. https://doi.org/10.1016/j.gloenvcha.2008.07.013

da Silva, J., Kernaghan, S., & Luque, A. (2012). A systems approach to meeting the challenges of urban climate change. *International Journal of Urban Sustainable Development, 4*(2), 125–145. https://doi.org/10.1080/19463138.2012.718279

Dewulf, A., Karpouzoglou, T., Warner, J., Wesselink, A., Mao, F., Vos, J., … Buytaert, W. (2019). The power to define resilience in social–hydrological systems: Toward a power-sensitive resilience framework. *WIREs Water, 6*(6), e1377. https://doi.org/10.1002/wat2.1377

Fast, V., & Rinner, C. (2014). A systems perspective on volunteered geographic information. *ISPRS International Journal of Geo-Information, 3*(4), 1278–1292. https://doi.org/10.3390/ijgi3041278

Fiksel, J. (2003). Designing resilient, sustainable systems. *Environmental Science & Technology, 37*(23), 5330–5339. https://doi.org/10.1021/es0344819

Findeisen, W., & Quade, E. S. (1997). The methodology of systems analysis: An introduction and overview. In H. J. Miser, & E. S. Quade (Eds.), *Handbook of systems analysis: Overview of uses, procedures, applications, and practice* (pp. 117–149). Chichester: John Wiley & Sons.

Ford, J. D., McDowell, G., Shirley, J., Pitre, M., Siewierski, R., Gough, W., … Statham, S. (2013). The dynamic multiscale nature of climate change vulnerability: An Inuit harvesting example. *Annals of the Association of American Geographers, 103*(5), 1193–1211. https://doi.org/10.1080/00045608.2013.776880

Frantzeskaki, N., Hölscher, K., Wittmayer, J. M., Avelino, F., & Bach, M. (2018). Transition management in and for cities: Introducing a new governance approach to address urban challenges. In N. Frantzeskaki, K. Hölscher, M. Bach, & F. Avelino (Eds.), *Co-creating sustainable urban futures: A primer on applying transition management in cities* (pp. 1–40). Cham: Springer.

Fraser, E. D. G., Dougill, A. J., Mabee, W. E., Reed, M., & McAlpine, P. (2006). Bottom up and top down: Analysis of participatory processes for sustainability indicator identification as a pathway to community empowerment and sustainable environmental management. *Journal of Environmental Management, 78*(2), 114–127. https://doi.org/10.1016/j.jenvman.2005.04.009

Gardner, J., Dowd, A.-M., Mason, C., & Ashworth, P. (2009). *A framework for stakeholder engagement on climate adaptation.* CSIRO Climate Adaptation National Research Flagship Working Paper 3. Retrieved from http://www.csiro.au/resources/CAF-working-papers.html.

Gramberger, M., Zellmer, K., Kok, K., & Metzger, M. J. (2015). Stakeholder integrated research (STIR): A new approach tested in climate change adaptation research. *Climatic Change, 128*(3–4), 201–214. https://doi.org/10.1007/s10584-014-1225-x

Hemstock, S. L., Des Combes, H. J., Buliruarua, L.-A., Maitava, K., Senikula, R., Smith, R., & Martin, T. (2018). Professionalising the 'resilience' sector in the Pacific Islands region: Formal education for capacity building. In S. Klepp, & R. Chavez-Rodrigues (Eds.), *A critical approach to climate change adaptation: Discourses, policies, and practices* (pp. 256–272). London: Routledge.

Hung, H.-C., Yang, C.-Y., Chien, C.-Y., & Liu, Y.-C. (2016). Building resilience: Mainstreaming community participation into integrated assessment of resilience to climatic hazards in metropolitan land use management. *Land Use Policy, 50,* 48−58. https://doi.org/10.1016/j.landusepol.2015.08.029

Hunt, A., & Watkiss, P. (2011). Climate change impacts and adaptation in cities: A review of the literature. *Climatic Change, 104*(1), 13−49. https://doi.org/10.1007/s10584-010-9975-6

Jha, A., Brecht, H., & Stanton-Geddes, Z. (2015). Building resilience to disasters and climate change in the age of urbanization. In I. Davis, K. Yanagisawa, & K. Georgieva (Eds.), *Disaster risk growth and livelihood: Investing in resilience and development* (pp. 7−27). London: Routledge.

Kaur, N., Steinbach, D., Agrawal, A., Manuel, C., Saigal, C., Panjiyar, A., … Norton, A. (2017). *Building resilience to climate change: MGNREGS and climate-induced droughts in Sikkim.* Retrieved from http://pubs.iied.org/pdfs/10188IIED.pdf.

Kernaghan, S., & da Silva, J. (2014). Initiating and sustaining action: Experiences building resilience to climate change in Asian cities. *Urban Climate, 7,* 47−63. https://doi.org/10.1016/j.uclim.2013.10.008

Landauer, M., Juhola, S., & Klein, J. (2019). The role of scale in integrating climate change adaptation and mitigation in cities. *Journal of Environmental Planning and Management, 62*(5), 741−765. https://doi.org/10.1080/09640568.2018.1430022

Lechner, S., Jacometti, J., McBean, G., & Mitchison, N. (2016). Resilience in a complex world—Avoiding cross-sector collapse. *International Journal of Disaster Risk Reduction, 19,* 84−91. https://doi.org/10.1016/j.ijdrr.2016.08.006

López-Marrero, T., & Tschakert, P. (2011). From theory to practice: Building more resilient communities in flood-prone areas. *Environment and Urbanization, 23*(1), 229−249. https://doi.org/10.1177/2F0956247810396055

Magnan, A. (2014). Avoiding maladaptation to climate change: Towards guiding principles. *SAPIENS, 7*(1). Retrieved from http://journals.openedition.org/sapiens/1680.

McCabe, A., & Halog, A. (2018). Exploring the potential of participatory systems thinking techniques in progressing SLCA. *The International Journal of Life Cycle Assessment, 23*(3), 739−750. https://doi.org/10.1007/s11367-016-1143-4

Mondino, E., Scolobig, A., Borga, M., & Di Baldassarre, G. (2020). The role of experience and different sources of knowledge in shaping flood risk awareness. *Water, 12*(8), 2130. https://doi.org/10.3390/w12082130

Moser, S. C., & Ekstrom, J. A. (2010). A framework to diagnose barriers to climate change adaptation. *Proceedings of the National Academy of Sciences of the United States of America, 107*(51), 22026−22031. https://doi.org/10.1073/pnas.1007887107

O'Donnell, E. C., Lamond, J. E., & Thorne, C. R. (2018). Learning and action alliance framework to facilitate stakeholder collaboration and social learning in urban flood risk management. *Environmental Science & Policy, 80,* 1−8. https://doi.org/10.1016/j.envsci.2017.10.013

Rasul, G., & Sharma, B. (2016). The nexus approach to water−energy−food security: An option for adaptation to climate change. *Climate Policy, 16*(6), 682−702. https://doi.org/10.1080/14693062.2015.1029865

Ribot, J. (2013). Vulnerability does not just fall from the sky: Toward multi-scale pro-poor climate policy. In M. R. Redclift, & M. Grasso (Eds.), *Handbook on climate change and human security* (pp. 164−172). Cheltenham: Edward Elgar. https://doi.org/10.4337/9780857939111.00016

Rodriguez, R. S., Ürge-Vorsatz, D., & Barau, A. S. (2018). Sustainable Development Goals and climate change adaptation in cities. *Nature Climate Change, 8*(3), 181−183. https://doi.org/10.1038/s41558-018-0098-9

Romero-Lankao, P., Gnatz, D. M., Wilhelmi, O., & Hayden, M. (2016). Urban sustainability and resilience: From theory to practice. *Sustainability, 8*(12), 1224. https://doi.org/10.3390/su8121224

Saunders, W. S. A., & Becker, J. S. (2015). A discussion of resilience and sustainability: Land use planning recovery from the Canterbury earthquake sequence, New Zealand. *International Journal of Disaster Risk Reduction, 14*, 73−81. https://doi.org/10.1016/j.ijdrr.2015.01.013

Schipper, E. L. F., Thomalla, F., Vulturius, G., Davis, M., & Johnson, K. (2016). Linking disaster risk reduction, climate change and development. *International Journal of Disaster Resilience in the Built Environment, 7*(2), 216−228. https://doi.org/10.1108/IJDRBE-03-2015-0014

Sharifi, A., Chelleri, L., Fox-Lent, C., Grafakos, S., Pathak, M., Olazabal, M., ... Yamagata, Y. (2017). Conceptualizing dimensions and characteristics of urban resilience: Insights from a co-design process. *Sustainability, 9*(6), 1032. https://doi.org/10.3390/su9061032

Star, J., Rowland, E. L., Black, M. E., Enquist, C. A. F., Garfin, G., Hoffman, C. H., ... Waple, A. M. (2016). Supporting adaptation decisions through scenario planning: Enabling the effective use of multiple methods. *Climate Risk Management, 13*, 88−94. https://doi.org/10.1016/j.crm.2016.08.001

Tompkins, E. L., & Adger, W. N. (2004). Does adaptive management of natural resources enhance resilience to climate change? *Ecology and Society, 9*(2). https://doi.org/10.5751/ES-00667-090210

Torabi, E., Dedekorkut-Howes, A., & Howes, M. (2018). Adapting or maladapting: Building resilience to climate-related disasters in coastal cities. *Cities, 72*, 295−309. https://doi.org/10.1016/j.cities.2017.09.008

Tyler, S., & Moench, M. (2012). A framework for urban climate resilience. *Climate and Development, 4*(4), 311−326. https://doi.org/10.1080/17565529.2012.745389

UNDP (United Nations Development Programme). (2010). *Designing climate change adaptation initiatives: A UNDP toolkit for practitioners.* Retrieved from https://www.adaptation-undp.org/resources/training-tools/designing-climate-change-adaptation-initiatives-toolkit-practitioners.

UNDP (United Nations Development Programme). (2020). *Human Development Report 2020. The next frontier: Human development and the Anthropocene.* Retrieved from http://hdr.undp.org/sites/default/files/hdr_2020_overview_english.pdf.

Viner, D., Ekstrom, M., Hulbert, M., Warner, N. K., Wreford, A., & Zommers, Z. (2020). Understanding the dynamic nature of risk in climate change assessments: A new starting point for discussion. *Atmospheric Science Letters, 21*(4), e958. https://doi.org/10.1002/asl.958

Williams, A., Kennedy, S., Philipp, F., & Whiteman, G. (2017). Systems thinking: A review of sustainability management research. *Journal of Cleaner Production, 148*, 866−881. https://doi.org/10.1016/j.jclepro.2017.02.002

Yokohata, T., Tanaka, K., Nishina, K., Takahashi, K., Emori, S., Kiguchi, M., ... Oki, T. (2019). Visualizing the interconnections among climate risks. *Earth's Future, 7*(2), 85−100. https://doi.org/10.1029/2018ef000945

7

Assessing urban resilience to cope with climate change

Maria Adriana Cardoso[1], Maria do Céu Almeida[1], Maria João Telhado[2], Marco Morais[2] and Rita Salgado Brito[1]

[1] *National Laboratory for Civil Engineering (LNEC), Lisbon, Portugal;*
[2] *Municipality of Lisbon (CML), Lisbon, Portugal*

1. Introducing urban-resilience assessment

Urban areas are dynamic, complex, and vulnerable systems involving multiple strategic urban services, stakeholders, and citizens. Strategic services essential to society include water and sewerage (water supply, and wastewater and stormwater management), waste management, energy supply, public lighting, transport, and public security. The potential effects of climate dynamics on the urban areas might lead to the aggravation of existing fragile conditions and the emergence of new hazards or risk factors, with major impacts on strategic urban services, people, the natural and built environment, and the economy. Climate-change challenges to urban areas thus require an integrated and sustainable approach to increase their resilience, with the allocation of the needed resources at all administrative levels. These include the development and implementation of disaster risk reduction (DRR) strategies, policies, plans, laws, and regulations in all relevant sectors (UNISDR, 2015).

This chapter describes a resilience-assessment framework to be used by cities and urban-service managers. The framework considers a structured and objective-driven assessment aiming at supporting the development and monitoring of cities' and urban services' resilience action plans, thus contributing to appropriate DRR investments. We then apply this assessment framework to the waste and mobility sectors in Lisbon (Portugal), identifying main opportunities to enhance their resilience.

Investing in Disaster Risk Reduction for Resilience. https://doi.org/10.1016/B978-0-12-818639-8.00003-X

1.1 Understanding urban-resilience assessment

Urban resilience refers, overall, to the ability of human settlements to withstand, recover quickly from, and adapt to any plausible hazards. Resilience to disruptive events comprises reducing the related risks and damage as well as the ability to rapidly bounce back to a stable state. Besides addressing DRR, resilience includes changes in circumstances (ARUP & Rockefeller Foundation, 2015; UN-Habitat, 2018a; UNDRR, 2017a, 2017b).

Climate-related events affecting the urban water cycle, such as heavy rainfall, seawater-level rising, or droughts, can result in direct impacts on cities' strategic services and cause cascading events (Evans, Djordjevic, Chen, & Prior, 2018; Russo, 2018). Public and private investment in disaster risk prevention and reduction through structural and nonstructural measures are essential to enhance the economic, social, health, and cultural resilience of persons, communities, countries, and their assets, as well as of the environment. Additionally, such measures are cost-effective and instrumental to save lives, prevent and reduce losses, and ensure effective recovery and rehabilitation after disasters (UNISDR, 2015).

A resilience-building framework focusing on the water sector should address all the phases of the urban water cycle and consider the interactions between urban services. Therefore, comprehensive and effective resilience planning requires an integrated holistic approach incorporating the interdependencies between systems and cascading effects (Vallejo & Mullan, 2017; see also Chapter 4, by Gotangco and Josol, in this volume). Assessing current and future resilience is a precondition for cities to know where they stand and to support decision-making on strategies, actions, and measures to adopt. It also provides the basis for planning in the long, medium, and short terms, including the development of resilience action plans and assessing their progress (Cardoso, Almeida, Telhado, Morais, & Brito, 2018; Sharifi, 2016).

Grounded on the United Nations' Agenda 2030 for Sustainable Development, several countries have adopted long-term international agendas that strongly contribute to boosting resilience building in cities, such as the Sendai Framework for Disaster Risk Reduction 2015−2030, the New Urban Agenda, and the Paris Agreement (Panda, 2019). All these agendas incorporate assessment steps for tracking their implementation (UNGA, 2016). The recognition of the relevance of resilience assessment resulted in the development of several dedicated tools and frameworks by a wide variety of stakeholders in different fields (ICLEI, 2010; ISO,

2017; Patel & Nosal, 2017; Rockefeller Foundation & ARUP, 2015; Sachs, Schmidt-Traub, Kroll, Durand-Delacre, & Teksoz, 2017; Summers, Smith, Harwell, & Buck, 2017; UN-Habitat, 2018b; UNDRR 2017a, 2017b; World Bank, 2015). These frameworks present substantial variations in their structure, content, and complexity according to their purposes (e.g., resilience evaluation, resilience planning, or progress monitoring), the tackled resilience scope (e.g., social, built environment, governance, or risk), and the focus (e.g., DRR, climate change, climate and energy, or cyber attacks) (Cardoso et al., 2020b).

For assessing urban resilience, it is important to take into account that cities are multidimensional entities and, therefore, urban resilience needs to consider multidisciplinary perspectives. Additionally, the resilience of a city is determined by diverse interacting systems and their relationships. For this reason, urban resilience also depends on the overall performance and capacity of its subsystems, not solely on the city's ability to cope with specific natural hazards or to adapt targeted areas to the impacts of climate change (Brugmann, 2012).

2. Resilience-assessment approach focused on the urban water cycle

Several organizations developed diverse resilience-assessment frameworks with different purposes, focusing on various themes and having distinct structures and formulations. When considering the urban water cycle, it is possible to identify several concomitant needs. To bridge them, we present in this chapter the development of a new resilience-assessment approach, mainly grounded on the frameworks proposed by UN-Habitat (2018b), Rockefeller Foundation and ARUP (2015), and UNDRR (2017a, 2017b). It primarily aims at adding to these the assessment of each specific strategic urban sector in isolation as well as the interactions among them (Cardoso et al., 2020b; Velasco et al., 2018). The framework was developed under the European RESCCUE Project—Resilience to Cope with Climate Change in Urban Areas—and targeted cities and urban-service managers as its users, considering a structured and objective-driven assessment. Its purpose is to support the development and monitoring of cities' and urban services' resilience action plans, therefore contributing to appropriate investments in disaster risk prevention and reduction.

The objective-driven RESCCUE framework developed for resilience assessment is grounded on and aligned with the "Disaster Resilience Scorecard for Cities" (UNDRR, 2017a, 2017b), developed by the United Nations Office for Disaster Risk Reduction. Besides Lisbon, it was applied to Barcelona (Spain) and Bristol (United Kingdom), as these three cities present diverse context characteristics and climate-change-related concerns. These applications not only tested the approach but also allowed us to identify some needs and opportunities to enhance resilience to climate change and to support prioritization of actions, contributing to strengthen and update city resilience action plans. The examples that we provide in this chapter illustrate the application of the framework to the waste and mobility sectors in Lisbon and contribute with adaptation strategies and measures to increase the city's resilience to climate change.

2.1 The RESCCUE framework for resilience assessment

The RESCCUE Resilience Assessment Framework (RAF) was developed considering climate-change prospects. It aims at supporting the development of resilience action plans with a focus on water, considering a multisectoral approach, and integrating different urban services. It focuses on the city and its services and infrastructures, being driven by resilience objectives.

The RAF (Cardoso et al., 2020b) is based on the following resilience dimensions defined by UN-Habitat (2018a, 2018b):
— organizational: having a focus on the city and integrating governance relations, all the involved services, and the involvement of the population;
— spatial: focusing on the city, the urban space, and the environment;
— functional: considering the strategic services in the city; and
— physical: addressing the urban assets/infrastructures.

The RAF was conceptualized in a tree-like structure. For each dimension, the RAF assigns specific resilience objectives, translating the ambitions to be achieved in the medium to long term by the city and the related services. Whereas the organizational and spatial dimensions are globally applied to the whole city, the functional and physical dimensions unfold into subdimensions, one for each service under assessment. For each objective, the framework specifies key criteria, expressing the different points of view through which the objectives will be assessed and identifying the related metrics. These metrics are parameters or functions to assess the criteria. For all the metrics, the RAF defined closed-ended questions, typically accompanied by a set of three to seven possible answers.

To facilitate the application of the RAF, each metric is assigned a relevance degree, which feeds the definition of the level of analysis (Cardoso, Brito, & Almeida, 2020a). The RAF considers three levels of analysis. The essential level includes all metrics with higher relevance required to integrate the resilience assessment of any city or service. On the other hand, the complementary level focuses on additional metrics to be considered whenever one seeks to integrate specific aspects of a city or service, hence corresponding to a more detailed resilience assessment. Finally, the comprehensive level considers additional metrics recommended whenever a more in-depth assessment is aimed at, for a city or service with a more mature resilience path. Conversely, depending on its resilience maturity, the city or service intending to apply the RAF may select a given set of metrics, according to their relevance.

The results of the framework's application to a city or service allow us to estimate its resilience-development level, per criterion and objective. We focus in this chapter on the functional- and physical-resilience assessment of Lisbon's urban services, specifically for the mobility and waste sectors.

2.2 RESCCUE functional- and physical-resilience dimensions to assess the mobility and waste sectors

We present a summary of the RAF's functional and physical dimensions in Table 7.1, considering the mobility and waste services.

The selection of the framework's application level for a specific city depends on the available information. Overall, the combination of easily obtainable data with available models and tools enables the analysis of urban resilience to climate change. The tasks to be undertaken within the framework's application depend on this combination. In general, the assessment of the services' functional and physical resilience considers the following tasks:

1. Characterization of the city, urban systems, and effects of climate variables;
2. Identification and assessment of potential climate-related hazards to urban services, for current and future scenarios, and identification of services' interdependencies;
3. Production of hazard maps to evaluate potential climate effects on strategic urban services and city resilience—for this analysis, a wide range of models may be used and can be integrated depending on the considered urban services and the available information and tools;

Table 7.1 Overview of the RAF regarding the functional and physical dimensions, for the mobility and waste services.

Functional dimension			Physical dimension		
Objectives	Criteria	Number of metrics Waste/Mobility	Objectives	Criteria	Number of metrics Waste/Mobility
	Strategic planning	5/5		Infrastructure assets' criticality and protection	5/4
	Resilience-engaged service	6/5	Safe infrastructure	Infrastructure assets robustness	12/11
Service planning and risk management	Risk management	9/7		Infrastructure assets' importance for and dependence on other services	4/3
	Reliable service	9/9			
	Flexible service	4/6	Autonomous and flexible infrastructure	Infrastructure assets' autonomy	6/1
	Service importance for the city	2/2		Infrastructure assets' redundancy	3/1
Autonomous service	Service interdependence with other climate-change services	2/2			
	Service preparedness for disaster response	4/0		Contribution to city resilience	4/3
Service preparedness	Service preparedness for climate change	8/6		Infrastructure assets' exposure to climate change	3/3
	Service preparedness for recovery and building back	13/0	Infrastructure preparedness	Preparedness for climate change	2/2
				Preparedness for recovery and building back	9/8

4. Assessment of exposure and vulnerability, considering interdependencies and cascading effects between the services; and
5. Assessment of functional and physical resilience focusing on city services and assets/infrastructures, respectively.

We illustrate herein the assessment of the functional- and physical-resilience dimensions for the mobility and waste sectors in Lisbon regarding the flooding hazard.

2.3 GIS-based analysis to support the assessment of mobility and waste sectors as regards flooding hazard

The steps considered for estimating the impact of flooding on city services, infrastructures, and users were threefold:
- The mapping of hazards (rainfall-induced and coastal-overtopping flooding);
- The mapping of infrastructures and components (roads, interfaces, and critical components, for traffic; zones and containers' location, for waste) classified according to their functional importance; and
- The analysis of spatial exposure and criticality.

Considering task 3 (see Section 2.2), in a data-limitation context, the rainfall-induced flooding-hazard maps can be obtained by processing historical records of flooding events, to proceed with detailed mathematical modeling of the drainage system. On the other hand, existing mathematical modeling tools of estuarine regimes enable obtaining coastal-overtopping flooding-hazard maps. In the Lisbon case, we considered different scenarios for both hazards.

For task 4, the approach that we adopted for the assessment of exposure and vulnerability of the mobility and waste services was a geographic information system (GIS)-based surrogate model, which uses available information and flooding-hazard maps. In this task, the RAF takes into account the interdependencies and cascading effects between services and permits obtaining results to support decision-making on possible courses of action to face the expected effects of climate dynamics.

The main outcomes in this step are urban areas' exposure for each analyzed scenario, herein illustrated for the mobility and waste sectors' critical components, and the estimation of broad impacts on the related service. Information from historical event records complement the hazard maps, namely regarding the range of water levels expected in different locations.

3. Functional- and physical-resilience assessment in Lisbon's mobility and waste sectors

3.1 General characterization

Starting with task 1, Lisbon is a municipality with a temperate climate and an extensive Tagus riverfront, shaped by a large number of cultural influences over time. It has a continental area of 85 km^2 with an average land slope of 5.7 degrees, reaching the maximum of 81 degrees, and an altitude ranging from 0 to 217 m. In 2011, the city had 547,733 residents, 24% of which older than 65 years, 13% of which younger than 15 years, and 17% of which disabled people (CML, 2016). The commuters' balance for travels between the place of residence and that of work or study is +378,226. A large number of tourists arrive annually in Lisbon—about 3 million tourists and 6.8 million tourist-nights per year. Regarding land use, 90% of the city falls in the consolidated-urban category, with approximately 52,500 buildings and 648,615 vehicles circulating per day (Vela et al., 2018). The city has around 80,000 water-distribution service connections, whereas 73% of its territory is served by a combined sewer network and three wastewater-treatment plants (Vela et al., 2018).

Climate-change trends for Lisbon include increase of the average air temperature, decrease of annual and nonwet-season rainfall, increase of wet-season rainfall and of the frequency of intense rainfall events, average-sea-level rise, and increase of the frequency of coastal floods (Vela et al., 2018). The combined action of intense rainfall, wind, and seawater-level rise with tides and storm surges is especially relevant for Lisbon's context and geographical position. The exposure and vulnerability vary according to the type of the affected system, namely its component and location, among other factors.

3.2 Identification and assessment of potential climate hazards

3.2.1 Water-related hazards, risk factors, and risks

For task 2, we carried out the identification and assessment of potential climate-related hazards affecting the urban services, for current and future scenarios, as well as the identification of services' interdependencies for water-related risks. Lisbon's main climate-change issues and challenges affecting its resilience are the increase of rainfall and associated risks and the increase of temperatures and droughts affecting water consumption. The aggravation of the variables disturbing the drainage-system performance (intense rainfall, average-sea-level rise, and wind intensity) has the potential to increase the risks associated with the overland flow-velocity level, flooding water level, and overflow contaminated with wastewater.

The deterioration of water quality in natural water bodies is relevant for the risks associated with recreational uses. Decreasing water quality derives from combined sewer overflows and reduced treatment effectiveness due to impacts on wastewater treatment plants (WWTP) from increased inflow, among others. The aggravation of the aforementioned variables also has the potential to increase equipment failure (including pumping facilities and WWTP equipment), as well as to limit the conveyance capacity of the networks (backwater effects). Moreover, higher temperature and salinity levels accelerate the degradation of pipes and mechanical equipment (Almeida, Brito, Cardoso, Beceiro, & Jorge, 2018; Póvoa, Pimentel, & Matos, 2018).

The major risks associated with Lisbon's drainage system are overland flows and flooding events, which have been relatively frequent in recent decades, with significant economic and social consequences (even if there were no direct fatalities). Regarding wastewater treatment, the most important variables are rainfall and seawater level, combined with tides and storm surges. If aggravated, these have the potential to increase the risk associated with excessive inflows and dilution, and the one of saltwater entry into the system leading to the reduction of the treatment's effectiveness (Linarić, Markić, & Sipos, 2013). Temperature is also a key variable affecting the performance of treatment processes, especially for biological ones. In Lisbon, the relevant hazards identified include flooding, windstorms, seawater level, storm surges, and heatwaves. For rainfall-induced events, flooding is frequent in Lisbon and one of the most relevant for the city's sectors.

Regarding mobility, the climate variables with potentially significant effects on transportation systems and services are rainfall, seawater level, storm surges, and wind. The exposure to water-related risks and disturbances is significant, and service interruptions influence a large number of societal functions. The interdependencies include increasing the likelihood of obstructions of urban-drainage components (inlets, trenches, and sewers) and negative effects on traffic (e.g., increase of the likelihood of traffic accidents and disturbance on circulation of vehicles and people as well as on economic activities). Main traffic routes are relevant for transportation within the city and essential for the commuters' vehicles, both on regular days and in emergencies. The comparison of recurrent flooding zones and transport networks and interfaces enhances the identification of overlaps and the potential for flooding affecting important circulation routes.

As for the waste sector, climate variables with potentially significant effects on its management services are rainfall and wind. For instance, flooding or intense wind can damage, displace, and overturn containers, potentially aggravating flooding levels. Temperature increase can deteriorate wastes at a faster rate, causing odor nuisances.

3.2.2 Interdependencies

The identification of interdependencies between urban services supports the analysis of the cascading effects in the city. Focusing on the waste and road-transport services and infrastructures in Lisbon, we present in Table 7.2 a summary of potential derived climate-related risks and cascading effects.

3.3 Flooding-hazard maps

Task 3 involves the production of hazard maps to evaluate potential climate effects on strategic urban services and city resilience. Herein, our focus is on flooding as an illustration of the proposed methodological approach. In Lisbon, we used a two-step procedure to produce the flooding-hazard maps and to estimate the potential effects of flooding-related climate pressures on strategic urban services. The results from the analysis of historical flooding data, meteorological data, estuary-level data, and modeling outcomes enabled to obtain a cross-validated mapping of flooding hazard (Fig. 7.1). The two main flooding-risk sources analyzed permitted to map the areas for flooding-frequency classes (for rain-induced flooding), as well as the inundated areas for specific time horizons (2050 and 2100, the latter is presented in Fig. 7.1).

3.4 Assessment of flooding exposure and vulnerability

3.4.1 Cascading effects: historical events

Task 4 consists in the assessment of exposure and vulnerability considering interdependencies and cascading effects between services. Lisbon has recorded several meteorological

Table 7.2 Summary of cascading effects for waste and road-transport services and infrastructures in Lisbon.

Service, subsystem, critical elements	Service or infrastructure failure	Variable/ hazard	Potential derived risks and cascading effects

Urban water cycle | Wastewater and rainwater systems

Wastewater treatment

— Wastewater treatment plant	— Excessive inflow and dilution generating lower treatment efficiency and combined sewer overflows (CSO) — Entry of saltwater into the system leading to potential corrosion of important infrastructures and causing lower treatment efficiency	Exposure to — Sea-level rise — High rain inflows, flooding, CSO	— Receiving water pollution — Recreational uses affected

Solid waste

Solid urban-waste collection (pneumatic collecting plants and network, waste vehicles)—not critical and not detailed

— Cleaning — Solid-waste containers	Damage, displacement, and overturn of containers	Exposure to — Rain/flooding — Wind/storms	Cascading effects considered on urban drainage: obstruction of components and surface flows; impacts on mobility

Solid urban waste treatment (waste treatment plant and cleaning stations)—not critical and not detailed

Transport | Roadways

— Main roads — Secondary roads — Local roads — Tunnels	Flooding and windstorm can cause interruption of public and private transportation (people and goods), including underground infrastructures	Exposure to — Sea-level rise — Rain/flooding — Wind/storms	Several urban services can be affected by cascading effects if maintenance or repair tasks are required during failures
— Traffic signals	Wind can generate failures of traffic-control systems	Exposure to — Rain/flooding — Wind/storms	Several urban services can be affected by cascading effects if maintenance or repair tasks are required during failures

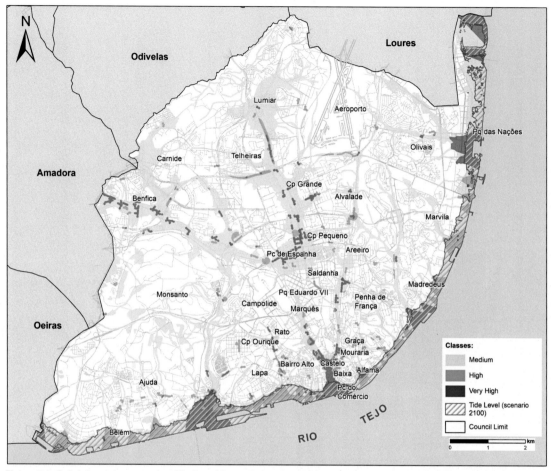

Figure 7.1 Rainfall- and estuary-level-induced hazard map for Lisbon—current situation, based on the Lisbon Master Plan (CML, 2012). Courtesy of CML-SMPC (2018).

events resulting in the occurrence of service failures with cascading effects on other city services or activities. The systematic analysis of these events enabled the identification of relevant interdependencies and the concerned services (Table 7.2). Intense rainfall, combined with local infrastructures' characteristics, results in insufficient conveyance capacity of the drainage systems, both overland and the underground system, generating high overland flows that reflect in higher water levels and flow velocities.

The main effects of these issues on mobility and transport infrastructures include flooding and damage of roads and underground infrastructures (metro, tunnels, and parking facilities), traffic and public-transport disturbances and interruptions, and damage to vehicles (Fig. 7.2). Regarding wastes, failures in the drainage system cause disruptions in the exposed waste-collection components. Flooding causes containers' damage through overturning, dragging, floating, and water retention. Overland flows can result in the spread of wastes on streets, and blockages of inlets and other drainage components. Therefore, spilled wastes can contribute to aggravating flooding (Fig. 7.3). Furthermore, the accumulation of debris on streets requires deep cleaning before resuming services.

Figure 7.2 Cascading effects from flooding and high estuary water level: Lisbon's mobility sector. Courtesy of Maria do Céu Almeida (2008) and CML (2008, 2014).

Figure 7.3 Cascading effects from flooding and high estuary water level: Lisbon's waste sector. Courtesy of CML.

The wastewater system's infrastructures, namely pumping stations and wastewater-treatment plants, can have cascading effects from failures in the drainage system, causing disruptions in these components' functions. They can also have secondary effects, for instance, in the Tagus Estuary, due to the discharge of wastewater with lower levels of treatment.

3.4.2 GIS-based analysis of the effects on mobility and road-transport infrastructures

The map of exposure and vulnerability to flooding considers the Lisbon municipality's road-network hierarchical classification based on road attributes, with five levels according to the Lisbon Master Plan (CML, 2012):

— First level: Structuring network ensuring intercounty connections, the crossing of the municipality, and the longest trips within the city;
— Second level: Main distribution network ensuring the distribution of the largest traffic flows toward the municipality, and the average routes and access to the structuring network;

- Third level: Secondary distribution network composed of internal routes, ensuring the distribution of proximity, and referral of traffic flows to the upper level; and
- Fourth and fifth levels: Local distribution/access network composed of structural routes (at the neighborhood level), in which the pedestrian has greater importance; these also include road accesses to buildings and the conditions for pedestrian circulation.

The lower levels correspond to the main routes, with lower redundancy, and consequently have higher vulnerability levels as interruptions at this scale have stronger effects on urban mobility. Daily commuters determine the traffic volume. Despite the limited data, an overview of the annual daily average flows led to the identification of critical locations at the city's boundaries. Additionally, we identified critical components for the functioning of the road network, namely tunnels and level crossings.

The main outcomes in this task consist in the analysis of the urban areas' exposure and transport sector's critical components for each scenario, together with the estimation of broad impacts on services. Information on historical events complemented the hazard maps, namely in the range of expected water levels at different locations. We also analyzed the effects of traffic infrastructures on hydrological processes that influence flooding risk in the city (Fig. 7.4). We carried out the calibration and validation of this surrogate-model approach by cross-checking its results with the recorded information related to past events, especially considering critical historical occurrences.

The results enabled the evaluation of the exposure of transport infrastructures and services for the different hazard levels. The vulnerability was evaluated based on the hierarchical classification of roads and interfaces. Taking into consideration the entire road network, less important roads in terms of traffic are more exposed to flooding: 33% of these are exposed to moderate to very high rainfall-induced flooding, and, globally, 41% of these roads' length is exposed. The road network is generally less exposed (5.3%) than the rail (27%) to estuary-level-induced flooding. An exploration of the interaction between potential effects of overland flows on the road-network infrastructure indicates that the impervious areas associated with Lisbon's road network represent about 12% of the municipal territory. The impact on the hydrological processes within the city, contributing to flooding, is thus significant.

3.4.3 GIS-based analysis of effects on waste services

We modeled the effects of flooding events on waste management in Lisbon considering recognized disruptive consequences

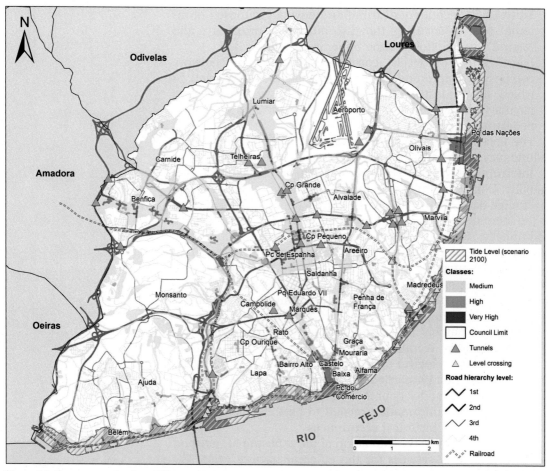

Figure 7.4 Exposure and vulnerability to rainfall- and estuary-level-induced flooding: Mapping for the transport networks. Courtesy of CML-SMPC (2018).

to this service and available data. Failure of the drainage system causes disruptions on waste-collection components, including containers' damage. Additionally, it can cause delays in collection operations. Overland flows can also result in the spillage of wastes on streets and the ensuing blockage of inlets.

The approach adopted herein took into account the different systems for waste collection in the city as a whole. The analysis of exposure and vulnerability to flooding considered the different solutions in place for waste collection. The main collection systems in Lisbon are: (1) large public containers (bring banks); (2) door-to-door (inside buildings, bins, or bags; outdoor only at specific days and hours); (3) pneumatic (network of underground vacuum conduits to collect wastes); and (4) self-delivering. The waste is collected by type in dedicated containers: mixed wastes, glass, paper, packages, biodegradable, batteries, and oils. Other wastes such as bulky scraps, furniture, garden litter, construction debris, and electrical equipment are collected on demand directly by trucks. Lisbon's territory is roughly divided into two parts in which the collection through the door-to-door system alternates. For all the solutions, there are established vehicle routes to transport the wastes to further processing.

We validated this surrogate-model approach by cross-checking its results with the recorded information about past events. Again, this step was undertaken especially for critical historical occurrences (Russo, 2018). Lisbon has 55,237 collection locations and 204,004 containers. The level of exposure of the waste-collection components to flooding is only 20% of the locations and 22% of the containers. The door-to-door system constitutes the predominant solution in terms of locations and containers (respectively, 64% and 82% of the total), but globally only about 20% of these are exposed to flooding.

The waste containers currently in use are already significantly resistant to flooding. The exposure and vulnerability to flooding effects are thus widely minimized with the types of solution utilized. We present the exposure of waste infrastructures for different hazard levels in Fig. 7.5. The vulnerability is reduced with the almost generalized use of mechanisms to avoid the movement of the containers that are permanently located in public spaces and with the door-to-door system largely implemented in the city. Often, the buildings' containers are protected inside the properties, being placed outdoor only a few hours per week, depending on the waste type.

3.5 Resilience assessment of Lisbon's mobility and waste sectors

Task 5 considers the functional- and physical-resilience assessment focusing on city services and assets/infrastructures, respectively. Here, we present the assessment results for the mobility and waste sectors as regards the flooding hazards.

3.5.1 Mobility sector

Using the RAF for the physical dimension of the mobility sector (Table 7.1), two different criteria permit assessing the objective of a safe mobility infrastructure. The first criterion is infrastructure

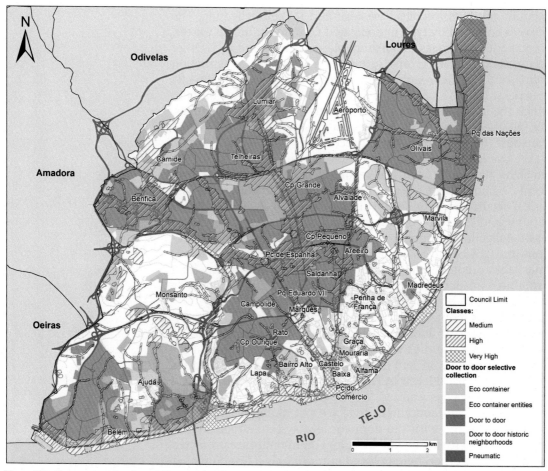

Figure 7.5 Exposure and vulnerability to rainfall- and estuary-level-induced flooding: Mapping for the waste collection system. Courtesy of CML-SMPC (2018).

assets' criticality and protection, which is measured by five metrics. In Table 7.3, we present this criterion and the possible answers for each metric, highlighting Lisbon's results in bold.

3.5.2 Waste sector

According to the RAF functional dimension for the waste service (Table 7.1), the flexibility of the service is one of the five criteria to assess the objective of having waste-service planning and risk management. The service's importance to the city is one of the two criteria to assess the objective of autonomous waste service. As presented in Table 7.4, these criteria may be

Table 7.3 Example of the application of the physical assessment to Lisbon's mobility/transport sector.

		Objective: Safe-mobility infrastructure			
Criterion: infrastructure assets' criticality and protection					
Metric	**Question**	**Answer (select)**			**Other**[a]
Mobility infrastructure's critical assets	Critical infrastructure assets for service provision are identified	**Yes**	Partially	No	
Component importance	Identification of infrastructure's critical assets is based on:	**Population served** Associated sensitive customers	**Location** High dependence on other services' infrastructures	Traffic during peak hours	Other services' infrastructures highly depend on mobility infrastructures
Mobility infrastructure's critical assets mapping, review, and update	Infrastructure's critical assets identified on hazard maps and included in risk data	Yes	**Partially, not covering all hazards or all infrastructure**[b]	No	
Exchange of information	Regular exchange of information with the city regarding infrastructure's critical assets, hazard maps, and risk data	**Yes, exchange of information from both sides**	The service is informed by the city	The city is informed by the service	No exchange
Protective buffers' mapping and information to the city	Protective buffers to safeguard infrastructure assets defined, clearly identified on hazard maps and risk data, and the city is informed	Yes	**Partially, or with a time horizon longer than ten years**	No	

[a] Based on functional importance and critical components for functioning: relevant tunnels and level crossings, commuting interfaces, type, volume, and passengers.
[b] Only for rain and tidal flooding.

Table 7.4 Example of the application of the functional assessment to Lisbon's waste sector.

Objective: Waste service planning and risk management				

Criterion: flexible service

Metric	Question	Answer (select)			
Treated solid waste recovered	% of treated solid waste recovered	Recycling and reuse 8	Energy recovery 70	Composting 7	Other
Solid-waste disposal	Solutions for solid-waste disposal in the city	Landfill	Incineration	Composting	Recycling
Solid-waste disposal location	Location of the city's solid-wastes disposal points	Within urban area	Outside city boundaries but within the metropolitan area		Far from the outskirts of the metropolitan area
Solid-waste service management	Adequate management (technological tools used, existing competencies, command chain)	Yes	No significant technological tools exist; competencies are adequate and a command chain is in place	Only a command chain is in place	No

Objective: autonomous waste service	

Criterion: service importance to the city

| Stakeholders' perception | Solid-waste service score based on stakeholders' perception | 2.94% | | |
| Cascading impacts | Understanding of potentially cascading failures between different services, under different scenarios | Yes | Cascading effects were studied, but climate-change scenarios were not considered | Cascading effects were studied only for some services; climate-change scenarios were not considered | No |

measured by four and two metrics, respectively. The RAF defined the set answers: for some metrics, a single choice needs to be made; for other cases, there are answers with multiple choices. Again, we present Lisbon's results in Table 7.4 in bold.

4. Appraisal of the functional- and physical-resilience results

4.1 Mobility sector

The results of the assessment of the functional and physical dimensions allowed us to identify how the city's mobility sector stands regarding its service provision and infrastructure, as well as the opportunities for improvements. Analyzing the objective of a safe-mobility infrastructure regarding the assets' criticality and protection (Table 7.3), we identified Lisbon's critical infrastructure assets for service provision based on the population served, location, and functional importance. The infrastructure's critical assets are only partially identified on hazard maps and included in risk data, namely concerning flooding. The city regularly exchanges information regarding critical assets, hazard maps, and risk data. Protective buffers to safeguard infrastructure assets are partially defined on hazard maps and in risk data.

The aspect acknowledged as requiring further development is the identification of critical assets so as to consider other services' infrastructures that are highly dependent on mobility infrastructures (such as stormwater), as their impact on the hydrological processes is significant. Another key aspect that needs to be further explored is the production of hazard maps and risk data considering hazards other than flooding, as well as the completion of the protective buffers' mapping and the provision of information to the city. Moreover, identifying the opportunities for improvement, the RAF supports the selection of measures to increase resilience in critical facilities, particularly through structural, nonstructural, and functional disaster risk prevention and reduction measures.

4.2 Waste sector

As regards the functional dimension of analyzing part of the objective of adequate waste-service planning and risk management (Table 7.4), Lisbon has a flexible service. This consists of diverse solutions for solid-waste treatment, recovery, and disposal, with disposal locations outside city boundaries, adequate service management with existing competencies and command chain, and the use of technological tools. Concerning the autonomy of the service, we identified opportunities for improvement related to cascading effects, which need to be studied for additional services and to consider climate-change scenarios, as well as to increase the public perception about this service.

For the physical dimension, as presented in the results ensuing from the GIS-based analysis (see Section 3.4.3), the level of flooding exposure and vulnerability of the components of the waste sector is low. On one hand, the containers are only placed outdoor a few hours per week. On the other, a significant number of the exposed containers already incorporate resilience measures, which provide them with high stability. Low exposure and vulnerability, together with the implementation of resilience measures, contribute to achieving the resilience objective related to waste-infrastruture preparedness (Table 7.1), through three criteria: climate-change exposure, climate-change preparedness, and preparedness for recovery and building back. The last criterion is assessed according to the RAF metrics "waste-collection infrastructure's component failures," "waste-management service locations unavailable," "waste containers damped or displaced," "level of failure of critical assets," and "time for restoration," regarding the last occurrence of a relevant climatic event.

Similarly, the assessment results of functional and physical dimensions were important to identify where the city's waste sector stands regarding its service provision and infrastructure, and also to spot opportunities for improvement. This can support the selection of measures and the definition of strategies to increase the resilience in the system, e.g., extending the implementation of stabilization or restraining mechanisms, or considering the option for underground alternatives. Like for the mobility sector, the RAF supports the selection of a wide range of DRR measures to increase resilience.

5. Final remarks on the resilience-assessment approach

The efforts to localize the Sendai Framework's third priority call for a better understanding of the particularities at the local level. In this regard, it is important to allocate the necessary resources, including finance and logistics, for the development and implementation of DRR strategies, policies, plans, laws, and regulations in all relevant sectors.

Assessing a city's current and future resilience constitutes the basis for diagnosis and decision-making regarding the adoption of the needed strategies, actions, and measures. It is also a key step for the development, implementation, and ex post evaluation of long-, medium-, and short-term plans. Comprehensive and effective resilience planning can benefit from integrated approaches incorporating interdependencies between systems. Following this rationale, we developed the RAF, an objective-driven framework for climate-change-resilience assessment with a focus on water, whose resilience objectives reflect the priorities of the Sendai Framework. The results of its application in three European cities emphasize the value of identifying needs and opportunities to enhance resilience to face climate-change impacts, thus supporting the prioritization of actions contributing toward cities' resilience action plans. The support provided in the selection of DRR measures to increase the resilience of critical facilities contributes to the allocation of the necessary resources for the development and implementation of DRR-related strategies and plans. It also provides a means to monitor and review progress after strategies' implementation.

The application of the RAF to Lisbon's waste and transport sectors—although focusing only on the functional- and physical-resilience dimensions—discloses key topics about the city's resilience building. For the mobility sector, particularly regarding a safe infrastructure with the identification of critical assets, the RAF enables us to identify the need to consider other services' infrastructures that highly depend on mobility infrastructures, as their impacts on hydrological processes are significant. It also shows the need to extend hazard mapping and risk data to other hazards. The assessment of the waste sector led to the conclusion that there are opportunities for improvement related to the autonomous-service objective, namely concerning cascading effects. Additionally, the low exposure and vulnerability of the assets, together with the resilience measures in place, contribute to achieving the infrastructure-preparedness objective.

The assessment approach presented in this chapter, combined with all the priorities of the Sendai Framework, the Paris Agreement on Climate Change, and the 2030 Agenda, highlights the relevance of cities' mission to reduce risks and become more resilient. In the case of Lisbon, an additional action level to further develop its resilience is to broaden the cooperation within the Lisbon Metropolitan Area. As illustrated in this chapter, the RAF can support cities and urban-service managers to localize the Sendai Framework's third priority toward better DRR investment decisions.

Acknowledgments

The RESCCUE Project was funded by the European Commission through Horizon 2020, the EU Framework Programme for Research and Innovation, under Grant Agreement number 700174. We would also like to thank Cristina Lucas Pereira, LNEC Research Fellow, for her support in this research.

References

Almeida, M. C., Brito, R. S., Cardoso, M. A., Beceiro, P., & Jorge, C. (2018). Approach to assess and control inflows into sewers. In J. Saldanha Matos, & M. João Rosa (Eds.), *Sanitation approaches and solutions and the sustainable development goals* (pp. 273–289). Lisbon: ERSAR, EWA, and APESB.

ARUP & Rockefeller Foundation. (2015). *City Resilience Framework*. Retrieved from https://www.rockefellerfoundation.org/wp-content/uploads/100RC-City-Resilience-Framework.pdf.

Brugmann, J. (2012). Financing the resilient city. *Environment and Urbanization, 24*(1), 215–232. https://doi.org/10.1177/0956247812437130

Cardoso, M. A., Almeida, M. C., Telhado, M. J., Morais, M., & Brito, R. S. (2018, November). *Assessing the contribution of climate change adaptation measures to build resilience in urban areas: Application to Lisbon*. Paper presented at the 8th International Conference on Building Resilience, Lisbon, Portugal.

Cardoso, M. A., Brito, R. S., & Almeida, M. C. (2020a). Approach to develop a climate change resilience assessment framework. *H₂Open Journal, 3*(1), 77–88. https://doi.org/10.2166/h2oj.2020.003

Cardoso, M. A., Brito, R. S., Pereira, C., Gonzalez, A., Stevens, J., & Telhado, M. J. (2020b). RAF Resilience Assessment Framework—A tool to support cities' action planning. *Sustainability, 12*, 2349. https://doi.org/10.3390/su12062349

CML (Câmara Municipal de Lisboa). (2012). *Relatório do Plano Diretor Municipal de Lisboa [Report of the Lisbon Master Plan]*. Lisbon: CML (in Portuguese).

CML (Câmara Municipal de Lisboa). (2016). *Relatório do estado do ordenamento do território 2015 [Stocktaking report on spatial planning 2015]* (Vol. I). Lisbon: CML/Departamento de Planeamento Urbano (in Portuguese).

Evans, B., Djordjevic, S., Chen, A. S., & Prior, A. (Eds.). (2018). *Development of methodologies for modelling of cascading effects and translating them into sectorial hazards*. RESCCUE Project Deliverable D3.3. Retrieved from https://toolkit.resccue.eu/wp-content/uploads/2020/12/d3.3._development_of_methodologies_for_modelling_of_cascading_effects_and_translating_them_into_sectorial_hazards.pdf.

ICLEI (Local Governments for Sustainability). (2010). *Changing climate, changing communities: Guide and workbook for municipal climate adaptation*. Retrieved from https://icleicanada.org/wp-content/uploads/2019/07/Guide.pdf.

ISO (International Organization for Standardization). (2017). *Sustainable development in communities—Inventory of existing guidelines and approaches on sustainable development and resilience in cities*. ISO 37121. Retrieved from https://www.iso.org/obp/ui/#iso:std:iso:tr:37121:ed-1:v1:en.

Linarić, M., Markić, M., & Sipos, L. (2013). High salinity wastewater treatment. *Water Science & Technology, 68*(6), 1400–1405. https://doi.org/10.2166/wst.2013.376

Panda, A. (2019). Foreword. In A. N. Martins, L. Hobeica, A. Hobeica, P. P. Santos, N. Eltinay, & J. M. Mendes (Eds.), *8th ICBR Lisbon Book of Papers* (p. 7). Lisbon: ULisboa/FAUL/CIAUD.

Patel, R., & Nosal, L. (2017). Defining the resilient city. United Nations University Centre for Policy Research's working paper 6. Tokyo: UNU.

Póvoa, P., Pimentel, N., & Matos, J. S. (2018). Assessment of salinity impacts in a wastewater system—the case of the Alcântara system in Lisbon, Portugal. In J. Saldanha Matos, & M. João Rosa (Eds.), *Sanitation approaches and solutions and the Sustainable Development Goals* (pp. 291–306). Lisbon: ERSAR, EWA, and APESB.

Rockefeller Foundation, & ARUP. (2015). *City Resilience Framework*. Retrieved from https://www.rockefellerfoundation.org/wp-content/uploads/City-Resilience-Framework-2015.pdf.

Russo, B. (Ed.). (2018). *Multi-hazards assessment related to water cycle extreme events for current scenario (public summary)*. RESCCUE Project Deliverable D2.4. Retrieved from https://toolkit.resccue.eu/wp-content/uploads/2020/12/d2.4._multi-hazards_assessment_related_to_water_cycle_extreme.pdf.

Sachs, J., Schmidt-Traub, G., Kroll, C., Durand-Delacre, D., & Teksoz, K. (2017). *SDG index and dashboards report 2017*. New York, NY: Bertelsmann Stiftung and SDSN.

Sharifi, A. (2016). A critical review of selected tools for assessing community resilience. *Ecological Indicators, 69*, 629–647. https://doi.org/10.1016/j.ecolind.2016.05.023

Summers, J. K., Smith, L. M., Harwell, L. C., & Buck, K. D. (2017). Conceptualizing holistic community resilience to climate events: Foundation for a climate resilience screening index. *GeoHealth, 1*, 151–164. https://doi.org/10.1002/2016GH000047

UN-Habitat. (2018a). *Resilience*. Retrieved from https://unhabitat.org/urban-themes/resilience.

UN-Habitat. (2018b). *City Resilience Profiling Programme: Guide to the City Resilience Profiling Tool*. Retrieved from http://urbanresiliencehub.org/wpcontent/uploads/2018/10/CRPT-Guide-Pages-Online.pdf.

UNDRR (United Nations Office for Disaster Risk Reduction). (2017a). *Disaster Resilience Scorecard for Cities: Preliminary level assessment*. Retrieved from https://www.unisdr.org/campaign/resilientcities/assets/toolkit/documents/UNDRR_Disaster%20resilience%20scorecard%20for%20cities_Preliminary_English_Jan2021.pdf.

UNDRR (United Nations Office for Disaster Risk Reduction). (2017b). *Disaster Resilience Scorecard for Cities: Detailed level assessment*. Retrieved from https://www.unisdr.org/campaign/resilientcities/assets/toolkit/documents/UNDRR_Disaster%20resilience%20scorecard%20for%20cities_Detailed_English_Jan2021.pdf.

UNGA (United Nations General Assembly). (2016). *Report of the open-ended intergovernmental expert working group on indicators and terminology relating to disaster risk reduction*. Retrieved from https://www.preventionweb.net/files/50683_oiewgreportenglish.pdf.

UNISDR (United Nations Office for Disaster Risk Reduction). (2015). *Sendai Framework for Disaster Risk Reduction 2015 – 2030*. Retrieved from https://www.preventionweb.net/files/43291_sendaiframeworkfordrren.pdf.

Vallejo, L., & Mullan, M. (2017). *Climate-resilient infrastructure: Getting the policies right*. OECD Environment Working Paper No. 121. Retrieved from https://www.oecd-ilibrary.org/docserver/02f74d61-en.pdf.

Vela, S., Almeida, M. C., Cardoso, M. A., Telhado, M., Coelho, L., Morais, M., … Cosco, C. (2018). *Identification of potential hazard for urban strategic services produced by extreme events*. RESCCUE Project Deliverable D2.1. Retrieved from https://toolkit.resccue.eu/wp-content/uploads/2020/12/d2.1_

identification_of_potential_hazard_for_urban_strategic_services_produced_
by_extreme_events.pdf.

Velasco, M., Russo, B., Martínez, M., Malgrat, P., Monjo, R., Djordjevic, S., …
Buskute, A. (2018). Resilience to cope with climate change in urban areas—A
multisectorial approach focusing on water—The RESCCUE Project. *Water,
10,* 1356. https://doi.org/10.3390/w10101356

World Bank. (2015). *City strength: Resilient Cities Program.* Retrieved from
https://openknowledge.worldbank.org/handle/10986/22470.

C

Building knowledge on disaster risk reduction investment

8

Incentives for retrofitting heritage buildings in New Zealand

Temitope Egbelakin[1], Olabode Ogunmakinde[2] and Sandra Carrasco[1]
[1]*School of Architecture and Built Environment, University of Newcastle, Newcastle, NSW, Australia;* [2]*Faculty of Society and Design, Bond University, Robina, QLD, Australia*

1. Background

Heritage buildings are valued structures within our society that play an important role in maintaining a country's history and culture, and are worthy of preservation. These structures are public goods and have become symbols of culture and identity, and hubs of individual and community life (Elsorady, 2014; Sesana, Gagnon, Bonazza, & Hughes, 2020). Moreover, heritage buildings contribute to economic development through tourism and leisure activities (Goded et al., 2017). For instance, in New Zealand, tourism, including visits to heritage sites and buildings, contributed NZD15.9 billion (about USD10.9 billion), equivalent to 6.1%, to the country's GDP in 2018, and represented 21% of foreign-exchange earnings (Statistics New Zealand, 2018).

Heritage buildings are characterized by historic construction techniques and materials that are subjected to degradation over time. In New Zealand, about 20,000 earthquakes with a magnitude above 2.5 occur every year (Earthquake Commission, 2017). About 160 heritage buildings were demolished during the 2010 and 2011 earthquakes in the eastern city of Christchurch (New Zealand Heritage Trust, 2012). Moreover, the country may lose over 8800 unreinforced-masonry heritage buildings that are at risk of earthquake destruction if not adequately strengthened. The vulnerability of these buildings threatens the history,

Investing in Disaster Risk Reduction for Resilience. https://doi.org/10.1016/B978-0-12-818639-8.00012-0

architecture, and culture of the people embedded in them. This is detrimental to community identity, local economies, livelihoods, and town-center viability in many suburban towns (Yakubu et al., 2017). As humans may not influence the seismic hazard, society can only work toward reducing the vulnerability of the building stock. Therefore, devising strategies to enhance the adoption of mitigation measures in heritage buildings is crucial for strengthening and preserving them.

Priority 3 of the Sendai Framework (UNISDR, 2015) provides guidelines for action at both local and national levels for investing in disaster risk prevention. This is to be achieved through the implementation of structural and nonstructural measures "to enhance the economic, social, health and cultural resilience of persons, communities, countries and their assets" (UNISDR, 2015, p. 18). Concerning heritage buildings, their owners can implement such measures to retrofit these properties. This would reduce the immediate risks of damage and loss during an earthquake and allow the acceleration of the recovery from postimpact conditions (Egbelakin, Wilkinson, Ingham, Potangaroa, & Sajoudi, 2017). Privately owned heritage buildings are characterized by voluntariness (i.e., engagement in any form of use under the law), exchange (i.e., choice to sell to someone else), inherent maintenance, and derived benefits. Yet, many private heritage-building owners are struggling to undertake adequate mitigation because of its associated high costs (Egbelakin, 2013). These owners often lack the motivation and financial justification for investing in the retrofitting and preservation of their heritage buildings. Egbelakin, Wilkinson, and Ingham (2014) observed that when heritage buildings become uneconomical to strengthen and preserve, their owners usually choose to abandon or demolish them.

Governments at all levels can motivate heritage-building owners through financial, regulatory, technological, and property market-based incentives (Egbelakin, 2013). Such incentives help ensure that building owners do not suffer economically from regulatory restrictions and can take advantage of capital investments in preservation (Mualam, 2015). According to Egbelakin et al. (2017), financial incentives, including tax deductions and insurance, are the most encouraging and enabling ones for preserving heritage structures. Nonfinancial incentives, such as warranties, are also used, for instance, in Japan (Fujimi & Tatano, 2013). Nevertheless, most of these incentives have had few successes, particularly for the owners of heritage buildings whose properties require a wide range of restoration and structural strengthening. In New Zealand, inadequate incentives could lead to the possible demolition of heritage buildings and unnecessary loss of valuable historic assets, especially in regions such as Auckland where the demand for land is high.

Meng and Gallagher (2012) observed that such incentives can be effective if properly designed and implemented, but several factors can affect their level of success. New Zealand provides financial and nonfinancial incentives to owners of heritage buildings for seismic retrofitting, restoration, and preservation works. Such incentives satisfy specific requirements, such as payment for professional services and materials, rapid approval process, and technical support during the retrofitting and preservation process. It is uncertain whether the current incentives in the country are adequate to motivate heritage-building owners to adopt risk-reduction measures. This study sets out to examine the preferences of building owners for incentives to retrofitting heritage structures in New Zealand. The objectives were (1) to examine heritage-building owners' level of awareness of available incentives and how it relates to their uptake in New Zealand; and (2) to identify the preferences of these owners for incentives to seismically strengthen and preserve heritage structures. This study intends to provide recommendations, including regarding potential incentive reforms, to local councils in New Zealand when considering investments for reducing the vulnerability of historic assets to earthquakes.

The rest of this chapter is structured as follows: In Sections 2 and 3, we review existing incentives around the world for heritage preservation. Section 4 presents the employed methodology, data sources, and analysis techniques. In Sections 5 and 6, we analyze the collected data and examine building owners' awareness and preferences for incentives in New Zealand. Finally, in Section 7, we conclude with the study's main findings and recommendations.

2. Factors affecting seismic retrofitting of heritage buildings

Seismic retrofitting implies strengthening heritage buildings that are vulnerable to earthquake impacts to become more resilient. Building owners are often reluctant to adopt this structural predisaster mitigation measure (McClean, 2012), despite its numerous advantages, including safer buildings, reduced loss of lives and property damage, and cost savings from emergency response and clean-up after a disaster. Identifying the factors affecting seismic retrofitting would be a significant step in encouraging heritage preservation. These factors are described in the following subsections.

2.1 Financial factors

Heritage-building owners are restricted by the costs associated with rehabilitating their earthquake-prone buildings, which may vary depending on the complexity and the amount of work needed to strengthen the structure (Nahkies, 2014). Egbelakin et al. (2014, 2015) and Fatorić and Biesbroek (2020) identified the cost of retrofitting and preserving or restoring existing materials in heritage buildings as a significant barrier to making these assets resilient. However, heritage-building owners' decisions are guided by the expected minimal returns (for instance, via rents) on making such high investments (Egbelakin, Wilkinson, & Ingham, 2015). Furthermore, Egbelakin et al. (2014) revealed that rental incomes and property price or valuation do not commensurate with retrofit costs, making retrofitting financially unviable.

Aside from the direct financial investment, these owners are reluctant to act due to the discrepancies between several policies associated with seismic retrofitting and heritage preservation, including those inherent to insurance and financial institutions (Ministry of Business, Innovation and Employment, 2013). According to Kunreuther and Michel-Kerjan (2009), high insurance premiums increase buildings' operational costs, making it hard for owners to undertake measures to mitigate seismic risks. The New Zealand Ministry of Business, Innovation and Employment (2016) suggested that the government creates appropriate insurance and financial instruments that would make seismic retrofitting easier for building owners. Market-based incentives such as investment tax, tradable permits, and user charges can motivate owners to adopt adequate mitigation measures in their heritage properties by offsetting the retrofitting cost.

2.2 Social factors

Egbelakin et al. (2015) emphasized the need for careful consideration of sociological factors for the successful mitigation of earthquake risks. According to MacGregor, Finucane, and Gonzalez-Caban (2008), these factors include risk perception, preconception about the relationship between higher seismic standards and retrofit costs, and a lack of trust associated with people, government agencies, and professional organizations.

For instance, some authors identified that the lack of trust in seismic and heritage-preservation techniques and professionals, as well as the difficulty in measuring the social and financial benefits from the heritage preservation and seismic retrofitting, affects building owners' mitigation decisions (Egbelakin et al., 2015; Egbelakin et al., 2017). Furthermore, discrepancies in the information on hazard mitigation often negatively affect people's decisions (Paton, 2007). This suggests the need for government agencies and industry experts to be consistent with the heritage preservation and earthquake-mitigation information they provide to building owners. It is therefore necessary for local councils to intervene by creating awareness about the importance and adequacy of predisaster risk-mitigation measures.

2.3 Regulatory factors

The New Zealand government enacts regulatory requirements for the seismic rehabilitation of heritage buildings, with technical recommendations from relevant industry organizations (Egbelakin, 2013). According to Egbelakin et al. (2014, 2015), factors such as minimum safety standards, as set by most building codes and policies, and a lack of a national mandate affect heritage preservation. Other factors include low levels of commitment by local councils to risk mitigation and the lack of political support for seismic retrofits and heritage community programs (Egbelakin et al., 2014, 2015). Conversely, Gizzi, Kam, and Porrini (2020) observed that in countries with high seismic-risk profile, amendments to the natural-hazard insurance legislation and minor changes to the council regulations could improve the risk-reduction efforts through differential premium rates and deductibles. Su and Li (2012) revealed that heritage-preservation efforts are often jeopardized by complicated management structures due to the participation of several stakeholders with varying perceptions. The lack of interface between conservation programs, community engagement, and urban development has an impact on the preservation of heritage structures, as identified by Badhreenath (2010). Similarly, Conejos, Langston, Chan, and Chew (2016) noted that regulatory requirements, such as disability access and energy efficiency, can impede heritage preservation, especially when considering adaptive reuse. Government intervention, in the form of financial and nonfinancial incentives, is thus necessary to motivate owners of heritage buildings to take seismic-retrofitting decisions and maintain their properties.

2.4 Technical factors

Some technical factors affect the seismic retrofitting of heritage buildings, as addressed in the literature. For example, Giaretton, Egbelakin, Ingham, and Dizhur (2018) highlighted that building owners and end users have concerns about the adoption of the technical solutions recommended by professionals, namely regarding high construction cost, building functionality and sustainability, esthetics, and poor restoration efforts. Perhavec, Rebolj, and Šuman (2015) observed that the lack of technical skills of craftsmen concerning older materials negatively affects the rehabilitation of heritage buildings. Similarly, Raheja (2013) reiterated that the lack of experience in traditional techniques equates to a low skill level in heritage preservation. Besides, the complexity of data management, hazard and risk assessments (La Rosa & Martinico, 2013), and anticipating material durability (Sanna, Atzeni, & Spanu, 2008) significantly affect the preservation and seismic retrofit of heritage structures.

Other deterrent factors include the use of uncommon materials (Erdem & Peraza, 2014), and the complexity of the design and implementation of seismic retrofitting in heritage buildings (Egbelakin, 2013; Giaretton et al., 2018). Each building should be approached based on its individual merits with distinct characters, hence posing a barrier for a one-size-fits-all technical solution. These factors indicate that the adequacy of technical skills and structural solutions is crucial to the success of seismic strengthening of heritage buildings. Therefore, there is a need to adopt a broad-ranging perspective to address these barriers to achieve meaningful impacts on heritage-building rehabilitation. The provision of adequate incentives can have a positive impact on owners' decision-making. In the next section we discuss some incentives available around the world and in New Zealand.

3. Incentives for retrofitting heritage buildings

Incentives can help to reduce the initial cost of strengthening heritage buildings by minimizing the pressures associated with project startup. Some countries in seismic-active regions such as North America, South East Asia, and Europe offer incentives to encourage owners of heritage buildings to invest in predisaster risk mitigation. Egbelakin (2013) identified four categories of such incentives, namely regulatory, financial, technical, and market-based ones. These incentives may circumvent the quandary of high retrofit costs and the difficulty of ensuring earthquake-resilient heritage buildings.

Nations of the world implement such incentives in different forms to motivate building owners. In the United States, the State of California provides financial incentives such as tax exemptions, grants, and low-interest loans to preserve heritage structures (Office of Historic Preservation, 2014). Some successful initiatives comprise regulatory incentives including relaxation of building regulations, such as flexible zoning and replacement of unavailable materials (Getty Conservation Institute, 2004). Also, in the United States and Australia, government agencies provide regulatory relief (density or floor-area-ratio bonuses and relief for nonconforming clauses) to building owners (Conejos et al., 2016). In South Korea, financial incentives (government subsidies, loans, tax relief, and planning) have been successful in helping to preserve heritage structures (Radzuan, Inho, & Ahmad, 2015). The UK government provides financial incentives (reduced tax and tax exemption) to building owners who comply with local heritage regulations (Araoz, 2008). Also, Pickerill and Pickard (2007) noted that property-tax rebate, a form of market-based incentive implemented in Ireland, was effective to encourage the preservation of heritage buildings. Similarly, the Japanese government offers heritage-building owners tax incentives, including related to national and municipal property taxes, to preserve their structures (Scott, 2006). Yet, the motivation of such building owners to intervene and hence increase the level of risk preparedness will depend on the appropriateness of the incentives.

In New Zealand, local and regional governments, as well as relevant organizations, provide financial incentives such as tax relief, fee waivers, loans, and grants to both public and private property owners (Egbelakin, 2013). For instance, the New Zealand Heritage Committee provides grants (up to NZD100,000, i.e., about USD69,000) to private owners (Department of Internal Affairs, 2018). The Wellington City Council introduced a three-year rate reduction, and the Built Heritage Incentive Fund has allocated NZD3 million (about USD2.1 million) over three years to encourage heritage preservation (Wellington City Council, 2018b). The charity organization Oamaru Whitestone Civic Trust established a financing scheme known as the Legacy Bond Fund for the preservation and retrofitting of heritage buildings (Oamaru Whitestone Civic Trust, 2018). Likewise, Heritage New Zealand provides financial incentives (up to NZD 100,000, i.e., about USD 69,000) primarily for the rehabilitation and maintenance of privately owned heritage buildings (Heritage New Zealand, 2018). Similarly, the councils of Christchurch and Matamata-Paiko offer revolving funds as a form of financial incentive, whereby the council buys, restores, and sells heritage buildings under special conditions, including maintenance requirements (Shipley, Utz, & Parsons, 2006). Some local councils, such as Auckland and Christchurch, also provide long-term low-interest loans to heritage-building owners. Such diversity suggests that, though financial incentives are available and accessible, they may not be sufficient to cover the cost of retrofitting heritage buildings. Considering the vast number of earthquake-prone heritage buildings in New Zealand, there is a need to explore different forms of incentives.

Furthermore, local and regional governments and Heritage New Zealand offer free technical assessment and guidance to heritage-building owners to promote heritage conservation and seismic retrofitting (Ross, 1997). They provide guidance on cost-efficient seismic-strengthening methods, databases of tradesmen who are skilled in certain historic materials, as well as advice on how and where these materials can be procured. The Wellington City Council created a Built Heritage Incentive Fund that contributes up to 15% of heritage-preservation costs, and a free initial engineering assessment (Wellington City Council, 2018a). Compared with other countries, New Zealand lags in its financial and policy incentives for encouraging the adoption of structural measures to reduce earthquake risks faced by its heritage buildings. There is also a lack of an inclusive community approach whereby everyone, including all government levels, communities, and building owners, has a role to play in promoting the rehabilitation and preservation of heritage assets.

Unfortunately, there are few incentive programs available for different classifications of heritage structures, particularly those privately owned. Whereas many private owners are unwilling to take a variety of structural risk-mitigation measures, the availability of incentives will encourage them to strengthen their buildings.

4. Research method

To achieve the research objectives, this study adopted a sequential mixed-method approach, comprising qualitative and quantitative techniques that complement each other (Creswell & Poth, 2017). The study focused on heritage-building owners in three regions in New Zealand: Invercargill, Gisborne, and Auckland. These regions were selected based on their high numbers of heritage buildings and their varying seismic-risk levels (Fig. 8.1), as established in the New Zealand Building Regulations (Ministry of Business, Innovation and Employment, 2016). Furthermore, in view of the different risk profiles of the selected regions, this study examined the causal relationships or potential correlations between risk zones and building owners' awareness of and preferences for incentives.

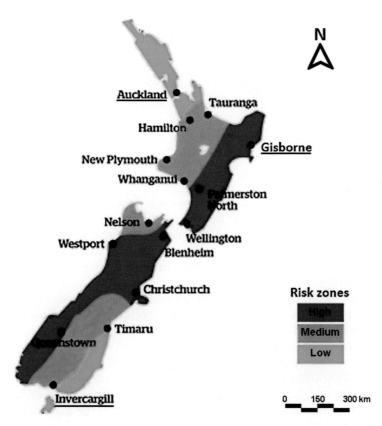

Figure 8.1 New Zealand's seismic risk zones. Adapted from Vaughan, O. (2017). *New quake laws mean property shake-ups*. Retrieved from http://truecommercial.oneroof.co.nz/insights/market-information/new-quake-laws-mean-property-shake-ups.

The participants were selected from two sampling frames. The first sample consisted of owners included in heritage-building databases maintained by the Invercargill, Gisborne, and Auckland councils. The second one comprised local council officials responsible for the design and implementation of the incentives. These officials could provide appropriate information, especially on the level of uptake of the incentives and the councils' capacity to provide them. The study only included officials with a minimum of three years of relevant experience.

The qualitative phase consisted of document analysis and semistructured interviews. Document analysis allowed for a review of existing incentives, policies, programs, reports, and legislation relating to seismic retrofitting and heritage preservation. Semistructured interviews provided insights, opinions, and explanations regarding the awareness, uptake, and preferences of available incentives. We conducted 21 interviews with 17 building owners and 4 council officials in the selected locations

Table 8.1 Distribution of the interviews and questionnaires.

Region	Number of heritage buildings	Number of interviewees	Number of valid questionnaires
Gisborne	261	7	16
Invercargill	84	8	26
Auckland	554	2	12

in 2018 (Table 8.1). The interviews comprised questions about the participants' awareness and knowledge of available incentives, preferences, and the mode of communicating these incentives to building owners. All the interviews were audiotaped, transcribed, and subsequently analyzed through thematic analysis with the help of the *NVivo* software.

In the quantitative phase, we developed a cross-sectional questionnaire based on the findings of the interviews to assess the preferences of building owners for the available and potential incentives for heritage preservation. We selected the respondents in the studied areas using a random-sampling technique. To capture their different opinions, we prepared a customized questionnaire comprising ten closed-ended questions measured on a five-point Likert scale. To identify possible biases in the questionnaire, we conducted a pilot test with ten industry practitioners and two scholars. We distributed 270 questionnaires to building owners randomly selected within the three regions via their email addresses. Out of these, 54 questionnaires were completed and used for the analysis, representing a response rate of 20%. We analyzed the preferences of building owners for incentives using descriptive statistics such as mean scores, ranking, and weighted average. To enhance data validity and reliability, we ensured that the interviewees and the questionnaire respondents differed. The following section presents the results of the interviews, whereas the findings of the survey are discussed in Section 6.

5. Awareness level and mode of communication of available incentives

Nine out of the 17 interviewees were unaware of the full range of incentives available to them both locally and nationally. We then asked the participants who were aware of available incentives, particularly in Gisborne and Invercargill, to specify them

and also to discuss the application process, if they were beneficiaries. Although none of the participants was aware of many of
the regulatory incentives, seven were aware of financial incentives, whereas one was familiar with technical assistance. The
need to use the incentives to offset the costs of retrofitting and
heritage preservation, as well as the high seismicity of Gisborne,
may be accountable for the high level of awareness in this region.
Moreover, this council has to address formal notices to owners to
take action on their seismic-prone heritage buildings. This is in
accordance with the national 2004 Building Act and the 2011
Gisborne earthquake policy, revised in 2016. Consequently, the
building owners in this city should be aware of the available
incentives for strengthening heritage properties. To substantiate
these findings, we sought to identify the councils' mode of
communication with heritage-building owners about the available incentives.

It was interesting to note that 9 out of the 17 participants were
unaware of some incentives offered within their regions. However, only 3 out of the 17 participants agreed that there could be
some incentives they are not aware of. Corroborating the importance of incentives for strengthening heritage buildings, 6 out of
the 17 participants agreed that increasing the current availability
of incentives would be beneficial, especially at the national level.
Consequently, when asked to specify the type of assistance they
may require, 7 out of the 8 participants who were aware of available incentives preferred financial assistance. Only one of them
preferred having appropriate technical support. This suggests
that financial assistance could motivate heritage-building owners
to undertake seismic retrofitting.

5.1 Gisborne

In this region, none of the participants showed interest in regulatory assistance. The perception that this type of incentive may
also serve as a form of punishment (Braithwaite, 2002; Kolieb,
2015) could be attributed to this result. The interviews showed
that despite Gisborne's seismic proneness, the owners were often
unaware of other available strengthening and preservation incentives outside their local jurisdiction. This lack of awareness can be
attributed to the low uptake of strengthening and preservation
activities. Therefore, it is plausible to conclude that the awareness
of the available incentives does not correlate to the adoption of
mitigation measures by building owners.

One of the reasons for the low awareness of available incentives in Gisborne could be the mode of communication about

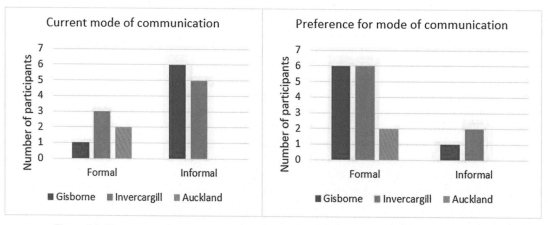

Figure 8.2 The contrast between current and preferred modes of communication. The authors.

them. Information is generally transmitted in two ways: formally, through official channels (including emails, letters, newsletters, conference proceedings, books, and reports), and informally, through remote or face-to-face discussions among friends and family. Of the seven participants in Gisborne, six expressed dissatisfaction with the way local authorities provide them with information. They thus rely on informal means of information but would prefer formal communication from councils through regular letters or emails, and meetings or workshops (see Fig. 8.2).

5.2 Invercargill

Half of the Invercargill participants were aware of the available incentives, which nonetheless have a low level of uptake. This may be attributed to the low value of financial incentives, the inadequacy of the information provided by the local council, and the region's medium seismicity. Of the four participants who were aware of available incentives, three were familiar with financial options, and one was conversant with technical-assistance incentives. None of the participants knew about the regulatory incentives. These findings indicate that the local council needs to raise awareness of all available incentives and encourage heritage-building owners to take advantage of them.

Two-thirds of the Invercargill participants revealed that there was no adequate communication from the council regarding the available incentives for the seismic strengthening of their heritage buildings after they were listed. Moreover, the uptake of incentives has been low, as revealed by a council representative: *"We have not received any applications for seismic strengthening of heritage buildings and corresponding incentives in the past one and a half years."* Based on this information, we infer that the awareness of the incentives in Invercargill may not influence the level of their uptake. This can be attributed to the councils' bottlenecks in the application processes for building approvals and securing the incentives, as revealed by the participants. Moreover, five Invercargill participants indicated that the incentives provided by their local councils were not adequate to address the costs of seismic retrofitting, and this may delay the preservation process. This puts in evidence the lack of appropriate communication between the building owners and their local councils. One of the participants mentioned that the current method of communication from the council is via letters delivered to mailboxes. Yet, such letters are often received by tenants, and when delivered to the building owners, the opportunity to act may have passed. Conversely, heritage-building owners in Invercargill prefer one-on-one communication, through which they could ask questions and clarify issues on the spot.

On the other hand, six participants in Invercargill indicated that incentives are important and, if adequate, could motivate them to strengthen and preserve their heritage buildings. When asked about the type of assistance they require, five out of eight participants preferred financial assistance (low-interest loans), two desired technical assistance (free information and advice services), whereas only one was interested in regulatory assistance (council's rapid planning process). These findings indicate that the financial incentives are of utmost importance to heritage-building owners in Invercargill, although their uptake is low.

5.3 Auckland

In view of the low level of seismic activity in the Auckland region, it was expected that most owners would be unaware of the incentives available for seismic strengthening. On the contrary, the two participants interviewed in Auckland were aware of the available government incentives. The level of awareness may be driven by market forces when building owners are motivated to seismically strengthen their buildings to benefit from high rental income in the region. The two participants claimed to be acquainted with the council's activities regarding heritage preservation, which justifies their high level of awareness. They were satisfied with the council's approach to sharing information through postal service (see Fig. 8.2). However, they suggested that an annual or biannual conference or workshop involving all heritage-building owners in Auckland would be a good approach to disseminate information and share experience. The two interviewed building owners demonstrated strong knowledge and understanding of seismic strengthening and heritage preservation by highlighting their advantages.

The fact that Auckland is the capital of New Zealand could lead to a better dissemination of information due to the presence of several government institutions. The two participants explicitly stated that they were aware of the financial incentives and could identify some of the requirements for financial incentives, which further confirms their level of awareness. They also reported that due to some ambiguity in the seismic-retrofitting policy, such as heritage-fabric protection and design provisions allowing for trade-offs between strength and ductility, they are yet to access the incentives to strengthen their buildings. The New Zealand building regulation (New Zealand Government, 2004) allows each local council to specify the timeframe to strengthen buildings based on the region's susceptibility to earthquakes. As emphasized by the participants, though there is a high level of awareness of available incentives within the Auckland region, their uptake is low due to ambiguity in local seismic-retrofitting policy. This may also be influenced by the low susceptibility of the Auckland region and the extended timeframe to act, as specified in the regulatory documents.

5.4 Aggregated results based on the interviews

The overarching view from these findings suggests that more than half of heritage-building owners interviewed in the three regions were not aware of the range of incentives available for seismic strengthening and heritage preservation. This was confirmed by one of the council officials who stated that heritage-building owners *"do not know about a lot of the options out there. Moreover, we have not been proactive about informing them."* Another council official observed that the incentives in New Zealand follow a specific pattern once established: *"We noticed that when a fund starts up, there is a bit of time necessary, about four years, for people to find out about it."* The councils must therefore be proactive in raising awareness about the incentives as they become available.

This lack of awareness indicates the high vulnerability of heritage buildings and the low level of risk preparedness for future seismic events. At the same time, the availability of incentives does not determine the level of adoption as indicated by the informants. Moreover, the provision of adequate financial incentives may be an effective means in motivating building owners to strengthen and preserve their properties. Yet, this can be a burden on the local councils, depending on the availability of funds for seismic retrofitting and heritage preservation. The unique circumstances—location, size, and taxpayer base—are important to the councils when providing and ensuring that property owners are aware of available incentives.

6. Building owners' preferences for incentives

In the three regions, we introduced and explained the incentives in detail to the building owners. The preferences of building owners for the available incentives for seismic retrofit and heritage preservation vary, as shown in Table 8.2. Based on the survey results, the most preferred incentive is flexibility to substitute building materials (R2). About 76% of the participants suggested that the council should maintain a list of potential material sources, vendors, and tradesmen who are skilled to use these specific materials.

Low-interest loan is the most preferred financial incentive for building owners, compared with the limited government financial assistance. This type of loan comes with little or no interest to minimize the financial burden of seismic-strengthening and heritage-preservation activities, and it allows building owners to negotiate affordable repayment terms. Although low-interest loans are a great way for building owners to finance their projects, the ensuing requirements should be flexible to reach small-business owners. For instance, such loans do not permit building owners to deposit in the loan account. Likewise, the process for borrowing additional money while servicing the current loan is strenuous. Most participants (89%) criticized the difficulty of accessing financial assistance, which suggests the need for the national government to provide policy instruments that enable access to such loans with less cumbersome processes.

As noted in the literature and confirmed by one of the interviewees, the lack of technical knowledge is a major challenge that does not encourage owners to rehabilitate their buildings. This may be addressed through free technical-assistance programs and advice on heritage preservation and seismic retrofitting, as identified by the respondents. Such assistance may include advice on

Table 8.2 Building owners' preferences for heritage-preservation incentives.

Code	Preference	Mean score	Rank
R2	Flexibility to use historic construction materials and methods alternatively to the prescribed regulatory standards	4.71	1
F11	Low-interest loans	4.60	2
T1	Free information and advice services	4.43	3
R7	Council's rapid planning process	4.40	4
R8	Heritage awards	4.33	5
F9	Funding for major repairs such as façades and roofs	4.32	6
F6	Low-income property owners' funds	4.28	7
F7	Funds from NZD8000 to NZD30,000 for the repair and redevelopment of foundations	4.20	8
F1	Tax-free revenue bonds to fund the renovation of heritage buildings	4.16	9
T2	Initial engineering and assessment report	4.16	9
F4	Tax credits in exchange for continued preservation of heritage buildings	4.14	11
F2	Tax credit of 10% or 20% on renovation investment in heritage buildings	4.00	12
F10	Subsidy on building-consent fees	4.00	12
M1	Property-tax relief	4.00	12
F8	Funds for senior-citizen property owners	3.83	15
F5	Funds to acquire, develop, and preserve heritage buildings	3.80	16
R4	Relaxation of construction requirements for listed heritage buildings	3.67	17
F3	Tax deduction on donation of part of the heritage building	3.60	18
R5	Relief for noncompliance with provisions	3.60	18
R6	Density and depth bonuses	3.60	18
R1	Reduction of development standards	3.50	21
R3	Parking relaxation requirements	3.00	22

innovative technical systems and materials, and demystifying regulatory requirements and approval processes. The councils' engineers and disaster experts may offer detailed explanations on the initial engineering-assessment report. These explanations could provide building owners with insights into the application requirements, processes, and procedures. Furthermore, they may boost the confidence of owners to undertake the rehabilitation of their buildings knowing that the council would support them.

Ranked next to free technical-assistance programs were councils' rapid planning and approval processes. As stated in the New Zealand Building Code, the processing time for building permits should be approximately 20 working days. The exemptions concern incomplete applications and complex structures. Due to their nature, heritage buildings can be categorized as complex structures that would require long planning and approval processes. As reported in this study, the planning and approval process can span between 8 and 24 months, depending on the building. Thus, about 82% of the owners feel that the council's rapid planning process for such buildings would motivate them to repair their properties. A participant suggested that the councils should give priority to the applications for seismic retrofitting and preservation considering the heritage buildings' vulnerability and location. Rapid planning and assessment processes can contribute to increased preparedness among property owners by providing adequate time for them to make early decisions.

Heritage awards also form part of the respondents' priority list. About 87% of the building owners wish to be appreciated and recognized for their efforts to preserve the country's rich cultural assets. Currently, Property Council New Zealand, through the Hawkins Heritage and Adaptive Reuses Property Award (Property Council New Zealand, 2018), acknowledges the achievements of building owners in preserving heritage structures. Although about 80% of the participants were not aware of this award, they strongly felt that rewarding their efforts would also motivate others.

7. Implications and summary of findings

This study identified and discussed the heritage-building owners' five most preferred incentives cut across their different categories. No single incentive category or policy tool could offer a broad-ranging solution; rather a combination of complementary means would produce the best result, necessary to motivate building owners to adopt structural risk mitigation measures in their properties. Consequently, policy-makers need to understand the owners' preferences and develop innovative strategies to address them. It is also clear from the study that building owners prefer financial incentives. Yet, the local councils indicated that the provision of financial incentives is unfeasible for them due to lack of funding. They may need to be funded through the generation of internal revenue, the establishment of endowment funds, and the prioritization of governmental allocations. This implies the need for national and local governments, as well as philanthropic organizations, to collectively promote disaster preparedness by providing financial incentives targeted at reducing the number of vulnerable buildings in New Zealand, thereby fostering protective actions against earthquake risk.

Furthermore, raising the awareness of available incentives for seismic retrofitting and heritage preservation is critical for their adoption. Although the local councils provide different types of financial, regulatory, market, and technical incentives, this study revealed that most building owners were unaware of their range. This study also showed that the awareness of the incentives does not relate to their uptake, but if well promoted these could motivate building owners to strengthen and rehabilitate their properties. Consequently, local councils across the country should increase their efforts to regularly disseminate these incentives. Heritage-building owners would feel more confident about adopting the incentives if they had been well informed. There is also the need to improve communication with building owners, not only by providing on-time and accurate information but also by devising effective means of consultation with them. Awareness raising and the adequacy of incentives could prompt owners to think and act accordingly, leading to the successful adoption of the Sendai Framework.

The availability of information, distributed via effective communication channels, is a hallmark of optimum risk preparedness. It is important that councils prioritize information dissemination and awareness activities. It is a fact that communication is complete when there are feedback loops. Although communication is full of complexities, it can succeed if language, values, and asymmetric-power issues are properly taken into consideration (see Chapter 5, by Buckman, in this volume). Therefore, councils should develop feedback systems that capture the opinions of building owners on the information they receive.

The restriction to only 3 out of the 16 regions of New Zealand and the size of the samples constitute the major limitations of this study. The inability of the study to equate awareness of incentives to motivation to act as regards seismic retrofitting is another shortcoming. Future studies may include more regions, particularly those recently affected by earthquakes. They may also examine the correlation between incentive awareness and motivation on a psychological and behavioral basis. Finally, further studies may evaluate the effectiveness and level of success of these incentives to motivate owners to undertake seismic retrofitting and preserve heritage buildings in New Zealand.

Acknowledgments

The authors would like to thank Ruan Malan, Rhiana McAnnalley, and Nitesh Suthar for their contribution to the data-collection process. We also appreciate the collaboration of the project participants from Gisborne, Invercargill, and Auckland.

References

Araoz, G. F. (2008). World-heritage historic urban landscapes: Defining and protecting authenticity. *APT Bulletin, 39*(2–3), 33–37.

Badhreenath, S. (2010). *Archaeological survey of India and its heritage conservation in Chennai Metropolitan Area*. Retrieved from www. cmdachennai.gov.in/pdfs/seminar_heritage_buildings/Archaeological_Survey_of_India_and_its_Hertiage_Conservation_in_CMA.pdf.

Braithwaite, J. (2002). Rewards and regulation. *Journal of Law and Society, 29*(1), 12–26. https://doi.org/10.1111/1467-6478.00209

Conejos, S., Langston, C., Chan, E. H., & Chew, M. Y. (2016). Governance of heritage buildings: Australian regulatory barriers to adaptive reuse. *Building Research & Information, 44*(5–6), 507–519. https://doi.org/10.1080/09613218.2016.1156951

Creswell, J. W., & Poth, C. N. (2017). *Qualitative inquiry and research design: Choosing among five approaches* (4th ed.). Los Angeles, CA: Sage.

Department of Internal Affairs. (2018). *Lottery Environment and Heritage Committee*. Retrieved from https://www.communitymatters.govt.nz/lotteryenvironment-and-heritage.

Earthquake Commission. (2017). *Earthquake statistics*. Retrieved from https://www.geonet.org.nz/about/earthquake/statistics.

Egbelakin, T. (2013). *Incentives and motivators for enhancing earthquake risk mitigation decision*. Doctoral dissertation. Auckland: University of Auckland. Retrieved from http://hdl.handle.net/2292/20133.

Egbelakin, T., Wilkinson, S., & Ingham, J. (2014). Economic impediments to successful seismic retrofitting decisions. *Structural Survey, 32*(5), 449–466. https://doi.org/10.1108/SS-01-2014-0002

Egbelakin, T., Wilkinson, S., & Ingham, J. (2015). Integrated framework for enhancing earthquake risk mitigation decisions. *International Journal of Construction Supply Chain Management, 5*(2), 34–51. https://doi.org/10.14424/ijcscm502015-34-51

Egbelakin, T., Wilkinson, S., Ingham, J., Potangaroa, R., & Sajoudi, M. (2017). Incentives and motivators for improving building resilience to earthquake disaster. *Natural Hazards Review, 18*(4), 04017008. https://doi.org/10.1061/(ASCE)NH.1527-6996.0000249

Elsorady, D. A. (2014). Assessment of the compatibility of new uses for heritage buildings: The example of Alexandria National Museum, Alexandria, Egypt. *Journal of Cultural Heritage, 15*(5), 511–521. https://doi.org/10.1016/j.culher.2013.10.011

Erdem, I., & Peraza, D. B. (2014). Challenges in renovation of vintage buildings. *Journal of Performance of Constructed Facilities, 29*(6), 04014166. https://doi.org/10.1061/(ASCE)CF.1943-5509.0000666

Fatorić, S., & Biesbroek, R. (2020). Adapting cultural heritage to climate change impacts in the Netherlands: Barriers, interdependencies, and strategies for overcoming them. *Climatic Change, 162*, 301–320. https://doi.org/10.1007/s10584-020-02831-1

Fujimi, T., & Tatano, H. (2013). Promoting seismic retrofit implementation through "nudge": Using warranty as a driver. *Risk Analysis, 33*(10), 1858–1883. https://doi.org/10.1111/risa.12086

Getty Conservation Institute. (2004). *Incentives for the preservation and rehabilitation of historic homes in the city of Los Angeles: A guidebook for homeowners*. Retrieved from http://www.getty.edu/conservation/publications_resources/pdf_publications/pdf/historic-homes.pdf.

Giaretton, M., Egbelakin, T. K., Ingham, J. M., & Dizhur, D. (2018). Multidisciplinary tool for evaluating strengthening designs for earthquake-prone buildings. *Earthquake Spectra, 34*(3), 1481−1496. https://doi.org/10.1193/091717EQS183M

Gizzi, F. T., Kam, J., & Porrini, D. (2020). Time windows of opportunities to fight earthquake under-insurance: Evidence from Google Trends. *Humanities and Social Sciences Communications, 7*, 61. https://doi.org/10.1057/s41599-020-0532-2

Goded, T., Beaupre, A., DeMarco, M., Dutra, T., Gogichaishvili, A., Haley, D., … Wright, K. (2017). Understanding different perspectives on the preservation of community and heritage buildings in the Wellington Region, New Zealand. *Natural Hazards, 87*(1), 185−212. https://doi.org/10.1007/s11069-017-2759-9

Heritage New Zealand. (2018). *National Heritage Preservation Incentive Fund.* Retrieved from http://www.heritage.org.nz/protecting-heritage/nationalheritage-preservation-incentive-fund.

Kolieb, J. (2015). When to punish, when to persuade and when to reward: Strengthening responsive regulation with the regulatory diamond. *Monash University Law Review, 41*(1), 136−162.

Kunreuther, H. C., & Michel-Kerjan, E. O. (2009). *Managing catastrophes through insurance: Challenges and opportunities for reducing future risks.* Working Paper #2009-11-30. Philadelphia, PA: Wharton Risk Management and Decision Processes Center.

La Rosa, D., & Martinico, F. (2013). Assessment of hazards and risks for landscape protection planning in Sicily. *Journal of Environmental Management, 127*, S155−S167. https://doi.org/10.1016/j.jenvman.2012.05.030

MacGregor, D. G., Finucane, M., & Gonzalez-Caban, A. (2008). The effects of risk perception and adaptation on health and safety interventions. In W. E. Martin, C. Raish, & B. Kent (Eds.), *Wildfire risk: Human perceptions and management implications, resources for the future press* (pp. 142−155). Washington, DC: RFF Press.

McClean, R. (2012). *Heritage buildings, earthquake strengthening and damage. The Canterbury earthquakes September 2010 − January 2012.* Retrieved from http://canterbury.royalcommission.govt.nz/documents-bykey/20120309.3756/$file/ENG.NZHPT.0004A.pdf.

Meng, X., & Gallagher, B. (2012). The impact of incentive mechanisms on project performance. *International Journal of Project Management, 30*(3), 352−362. https://doi.org/10.1016/j.ijproman.2011.08.006

Ministry of Business, Innovation and Employment. (2013). *Building seismic performance proposals to improve the New Zealand earthquake-prone building system.* Retrieved from http://www.mbie.govt.nz/info-services/buildingconstruction/documents-and-images-library/safety-quality-epb/building-seismic-performanceconsultation-summary-of-submissions.pdf.

Ministry of Business, Innovation and Employment. (2016). *Building code and handbooks.* Retrieved from https://www.building.govt.nz/building-code-compliance/building-code-and-handbooks.

Mualam, N. Y. (2015). New trajectories in historic preservation: The rise of built-heritage protection in Israel. *Journal of Urban Affairs, 37*(5), 620−642. https://doi.org/10.1111/juaf.12168

Nahkies, P. B. (2014, January). Mandatory seismic retrofitting: A case study of the land use impacts on a small provincial town. *Paper presented at the 20th Annual Pacific-Rim Real Estate Society Conference, Christchurch, New Zealand.* Retrieved from http://www.prres.net/papers/Nahkies_Mandatory_Seismic_Retrofitting_A_Case_Study.pdf.

New Zealand Government. (2004). *Building Act.* August 24, 2004.

New Zealand Heritage Trust. (2012). *Adapt and survive: New life for old buildings*. Auckland: Heritage New Zealand.

Oamaru Whitestone Civic Trust. (2018). *Legacy Building Fund—Preserving and revitalising Oamaru's Victorian Precinct*. Retrieved from https://www.victorianoamaru.co.nz/legacy-building-fund.

Office of Historic Preservation. (2014). *Incentives and grants for historic preservation*. Retrieved from http://ohp.parks.ca.gov/?page_id=1073.

Paton, D. (2007). Preparing for natural hazards: The role of community trust. *Disaster Prevention and Management, 16*(3), 370–379. https://doi.org/10.1108/09653560710758323

Perhavec, D. D., Rebolj, D., & Šuman, N. (2015). Systematic approach for sustainable conservation. *Journal of Cultural Heritage, 16*(1), 81–87. https://doi.org/10.1016/j.culher.2014.01.004

Pickerill, T., & Pickard, R. (2007). A review of fiscal measures to benefit heritage conservation. *RICS Research Paper Series, 7*(6). Retrieved from https://arrow.tudublin.ie/cgi/viewcontent.cgi?article=1011&context=beschrecart.

Property Council New Zealand. (2018). *Rider Levett Bucknall Property Industry Awards 2018: Call for nominations*. Retrieved from https://www.propertynz.co.nz/sites/default/files/uploaded-content/websitecontent/Forms/call_noms_natawards2018.pdf.

Radzuan, I. S. M., Inho, S., & Ahmad, Y. (2015). A rethink of the incentives programme in the conservation of South Korea's historic villages. *Journal of Cultural Heritage Management and Sustainable Development, 5*(2), 176–201.

Raheja, G. (2013). *Sandhi notes*. Retrieved from https://www.iitsystem.ac.in/sites/default/files/static/IIT-Council/mhrdinitiatives/41d30c21.pdf.

Ross, C. (1997). *Incentives for heritage conservation: The role of local and regional government in New Zealand*. Wellington: New Zealand Historic Places Trust.

Sanna, U., Atzeni, C., & Spanu, N. (2008). A fuzzy number ranking in project selection for cultural heritage sites. *Journal of Cultural Heritage, 9*(3), 311–316. https://doi.org/10.1016/j.culher.2007.12.004

Scott, G. R. (2006). A comparative view of copyright as cultural property in Japan and the United States. *Temple International and Comparative Law Journal, 20*(2), 284–362.

Sesana, E., Gagnon, A. S., Bonazza, A., & Hughes, J. J. (2020). An integrated approach for assessing the vulnerability of World Heritage Sites to climate change impacts. *Journal of Cultural Heritage, 41*, 211–224. https://doi.org/10.1016/j.culher.2019.06.013

Shipley, R., Utz, S., & Parsons, M. (2006). Does adaptive reuse pay? A study of the business of building renovation in Ontario, Canada. *International Journal of Heritage Studies, 12*(6), 505–520. https://doi.org/10.1080/13527250600940181

Statistics New Zealand. (2018). *Tourism satellite account: 2018*. Retrieved from https://www.stats.govt.nz/information-releases/tourism-satellite-account-2018.

Su, M., & Li, B. (2012). Resource management at world heritage sites in China. *Procedia Environmental Sciences, 12*(A), 293–297. https://doi.org/10.1016/j.proenv.2012.01.280

UNISDR (United Nations International Strategy for Disaster Reduction). (2015). *Sendai Framework for Disaster Risk Reduction 2015–2030*. Retrieved from http://www.wcdrr.org/uploads/Sendai_Framework_for_Disaster_Risk_Reduction_2015-2030.pdf.

Wellington City Council. (2018a). *Built Heritage Incentive Fund*. Retrieved from https://wellington.govt.nz/services/community-andculture/funding/council-funds/built-heritage-incentive-fund.

Wellington City Council. (2018b). *Incentives to strengthen earthquake-prone buildings*. Retrieved from https://wellington.govt.nz/services/rates-andproperty/earthquake-prone-buildings/help-for-earthquake-prone-building-owners/incentives-tostrengthen-earthquake-prone-buildings.

Yakubu, I. E., Egbelakin, T., Dizhur, D., Ingham, J., Sungho Park, K., & Phipps, R. (2017). Why are older inner-city buildings vacant? Implications for town centre regeneration. *Journal of Urban Regeneration and Renewal, 11*(1), 44–59.

9

Dissatisfaction after postdisaster resettlement

Pournima Sridarran[1], Kaushal Keraminiyage[2] and Dilanthi Amaratunga[3]

[1]*University of Moratuwa, Moratuwa, Sri Lanka;* [2]*University of Salford, Salford, United Kingdom;* [3]*Global Disaster Resilience Centre, University of Huddersfield, Huddersfield, United Kingdom*

1. Criticisms on postdisaster resettlement

Disasters triggered by hazards such as earthquakes, windstorms, landslides, floods, and tsunamis, as well as conflicts, are among the well-known causes that generate and aggravate homelessness (Dikmen & Elias-Ozkan, 2016). This is a critical issue for countries with poor economic stability. Despite the involvement of many governments and other relevant national and international entities to tackle disaster-induced displacements at different scales, many extensive resettlements show failure signs. These are often criticized for only being able to facilitate temporary relief without ensuring the long-term expectations of the affected communities.

Resettlement is a process of introducing a new built environment to a displaced population. This potentially redraws the social system as this process is interlinked with all other subsystems of the related community. Considering this complex nature, it is fair to assume that the new built environment introduced to the displaced community may disturb preexisting relationships and induce new vulnerabilities (both physical and social). Thus, the role of the built environment within the recovery process of displaced communities is an important topic of investigation that requires the attention of experts from various disciplines.

Betts (2009) argued that disaster-induced resettlements involve significant political interventions that often undermine the choice of the community to remain. In fact, the relocation of communities in such instances frequently follows a top-down

Investing in Disaster Risk Reduction for Resilience. https://doi.org/10.1016/B978-0-12-818639-8.00014-4

decision taken by the government or other responsible authorities. The resettlement hence becomes an involuntary act. Particularly in developing countries, governments assume the responsibility for such initiatives (Ganapati, 2016). To endorse this, the Sendai Framework calls for prioritizing "the adoption of policies and programmes addressing disaster-induced human mobility to strengthen the resilience of affected people and that of host communities, in accordance with national laws and circumstances" (UNISDR, 2015, p. 20). This is one of the key actions at the national and local levels under the Framework's Priority 3, in recognition of the global growth in the numbers of displacements and resettlements.

On the other hand, several studies draw attention to enduring criticisms of large-scale resettlements for failure to harmonize with the communities in the long term (Andrew, Arlikatti, Long, & Kendra, 2013; Barenstein & Leemann, 2012; Gunawardena & Wickramasinghe, 2009; Vidal López, 2011). Resettlement programs, particularly those that follow a donor-driven approach, are often criticized for their inability to provide the desired level of adaptability for both the built environment and the concerned people. Ganapati (2016) considered this a resettlement failure under two categories. Firstly, the author identified project-related failures, such as poor planning, implementation, coordination, and participation. Secondly, there are outcome-related failures, such as culturally unsuitable houses, the use of inappropriate materials and technology, and eventually the inability to meet the needs and expectations of the community. Furthermore, Ahmed (2011) pointed to organizational reasons for the shortcomings of donor-driven resettlements: the lack of institutional coordination, planning, and clear policy; inequitable distribution, corruption, inordinate construction delays; and financial mismanagement and misappropriation. The abovementioned arguments raise the question: Why do postdisaster resettlements often end up in dissatisfaction? This chapter aims to bring answers to this question by examining the views of resettled communities.

2. Disaster-induced displacement and resettlement

Over time, the term "disaster" has undergone a paradigm shift, from previously describing a war to currently characterizing social vulnerability and state of uncertainty (Quarantelli, 1995). In the late 1990s, Combs, Quenemoen, Parrish, and Davis (1999, p. 1125) defined a disaster as "a time- and place-specific event that originates in the natural environment and the resulting disruption of the usual functions and behaviors of the exposed human population." This definition explains disaster as simply driven by a natural event and fails to recognize that natural hazards become risks—and possibly materialize into disasters—due to human action (or inaction). Furthermore, it does not reflect the severity of the disruption. Later in 2009, the United Nations International Strategy for Disaster Reduction extended this definition as "a serious disruption of the functioning of a community or a society involving widespread human, material, economic or environmental losses and impacts, which exceeds the ability of the affected community or society to cope using its own resources" (UNISDR, 2009, p. 9). This definition implies that a disaster is any event, driven by natural or human-made hazards, that causes disruption and puts a community in a situation in which external resources are needed for recovery. The more recent definition hence includes a wide range of damaging events and clearly identifies their severity, though it does not reflect disasters' indirect impacts.

Based on this definition, disasters may be classified into five major categories: sudden-impact disasters, slow-onset disasters, epidemic disasters, industrial/technological disasters, and complex emergencies (Robinson, 2003). This classification highlights the timescale associated with each disaster and the ensuing displacement, and provides an insight for better planning and implementation of resettlements. Accordingly, those categories can be regrouped under two broad types, namely disasters inducing immediate displacements and disasters causing potential displacements. Sudden-impact disasters and complex emergencies are location-specific and trigger immediate displacements, whereas epidemics and industrial/technological disasters may induce either immediate or eventual displacements depending on the severity of the event. Slow-onset disasters may cause resettlements and seldom displacements. In this chapter, with the term disaster-induced displacement and resettlement (DIDR), we consider both conflict-induced and hazard-induced displacements and resettlements.

3. Dimensions of the postdisaster cycle

The characterization of the postdisaster cycle as a four-phase process is well established in the disaster-related research (Berke, Kartez, & Wenger, 1993; Joshi & Nishimura, 2016). However, it is often implied in the literature that these phases—emergency response and relief, recovery and reconstruction, mitigation, and preparedness—are sequential and expect to follow the given order (Oloruntoba, Sridharan, & Davison, 2018). This idea is challenged by Cronstedt (2002), who argued that such implicit sequential nature creates an artificial barrier between the postdisaster phases and leads to an "over the wall" treatment (Fig. 9.1). Although he agreed with the fact that disaster officials can separately identify the characteristics of each of these phases, this author argued that they are interconnected and overlap with each other. Therefore, these four phases should be treated appropriately when implementing disaster-response plans, without neglecting their interconnected nature (Jahre, Persson, Kovács, & Spens, 2007). Thus, for instance, the preparedness for future disasters should be developed during recovery and reconstruction.

The authorities in charge of postdisaster planning and implementation should hence recognize the multifaceted nature of building back better in their work. Moreover, the four phases of the postdisaster cycle involve different tasks whose weights depend on political decisions, including the prohibition of reconstruction in any specific landscape. In such situations, the recovery and reconstruction phase is particularly critical as it entails the creation of an entirely new built environment for the displaced population. The criticality further varies depending on who takes the responsibility for each phase.

Within the current literature on postdisaster resettlements, the success factors for such processes are only vaguely discussed. However, some related studies often cite the beneficiaries' dissatisfaction as one of the major shortcomings of large-scale postdisaster resettlement programs (Barenstein, 2015; Dikmen & Elias-Ozkan, 2016). In more specific terms, many authors argued that such dissatisfactions in many resettlement schemes

Figure 9.1 The four phases of the postdisaster cycle separated by artificial barriers. The authors.

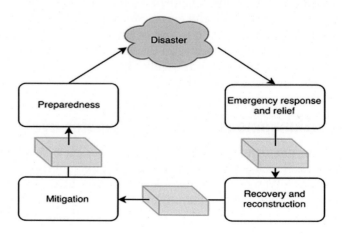

around the world are linked with issues related to the built environment (Jigyasu & Upadhyay, 2016; Muggah, 2008; Oliver-Smith, 1991).

4. Factors affecting the adaptation of resettled communities to a new environment

The term "built environment" refers to buildings and all the human-made structures, including uncovered settings, such as compounds, sites, and landmarks, which define the open spaces that support human activities (Lawrence & Low, 1990). The relationship between the built environment and the patterns of human activity is defined differently by two schools of thought. The first considers that the built environment adapts to the patterns of human activities (Jigyasu & Upadhyay, 2016), whereas the second believes that human activities adapt to the built environment (Barenstein, 2015). However, anthropological studies posit that both the built environment and human activities adjust to each other (Lawrence & Low, 1990).

Irrespective of which of these statements is more accurate, the adaptation to a new environment cannot be expected in the short term without ensuring certain features that facilitate basic human needs such as food, shelter, sanitation, and socialization. Whereas these fundamental human needs define the short-term adaptability, the long-term satisfaction of resettled communities within a new environment is a complex phenomenon. Based on previous studies (Andrew et al., 2013; Barenstein & Leemann, 2012; Foresight, 2011; Gunawardena & Wickramasinghe, 2009), we grouped the long-term struggles in adapting to a new built environment after disasters under three broad categories: housing, infrastructure, and location.

4.1 Housing

Disasters strengthen preestablished practices of social discrimination by weakening the most vulnerable populations (Cannon, 2008). It is not by chance that developing countries account for about 98% of the total disaster-induced homelessness (Gilbert, 2001). Housing reconstruction thus constitutes a key component of any postdisaster recovery or resettlement initiative, particularly in these countries. Furthermore, people often consider houses a highly valuable asset, thereby making these a prioritized need after a disaster (Ahmed, 2011). Despite this urgency, housing construction typically takes long (Gilbert, 2001). Hence, governments and donor agencies tend to seek expedite construction by taking quick decisions. Jigyasu and Upadhyay (2016) pointed out that such hastiness creates struggles for the displaced populations in adjusting to their new built environments.

4.2 Infrastructure

The essential human-made installations (henceforth referred to as infrastructures) that support the functionality of buildings and communities are also a key aspect of any built environment (Moteff & Parfomak, 2004). Generally, in a postdisaster resettlement operation, a large portion of the resettlement funds is directed to housing construction (Freeman, 2004). Consequently, the investments in infrastructures are often less than desirable. Postdisaster shortcomings related to insufficient infrastructures are highlighted in many studies (Barenstein, 2015; Gunawardena & Wickramasinghe, 2009; Seneviratne, Amaratunga, & Haigh, 2016). In addition, other studies show that the available resources in many host environments become also overwhelmed due to the lack of adequate infrastructure support (IFRC, 2013). Hence, besides being a problem for displaced communities, the lack of infrastructures can also degrade the performance of host communities' built environments.

4.3 Location

By determining the degrees of accessibility and maintenance of previous social and spatial linkages, the location of the resettlements plays a vital role in the adaptation to the built environment. It is true that resettlement processes cannot simply follow a complete bottom-up approach, in which the resettled community is empowered to participate in the decision-making process. For example, the selection of the location often depends on several external factors that are beyond the control of the involved communities. These include land availability, its capability to accommodate large-scale construction, and its susceptibility to future disasters. Such technical constraints may lead the institutions in charge of resettlement processes to expect that people somehow become accustomed to their new location (Grundy-Warr & Rigg, 2016; Staupe-Delgado, 2020). However, in such selections, the suitability of the land for the resettled communities' livelihood and other activities is often overlooked or paid least attention to (Gong, Zhang, Yao, Liu, & Wang, 2020; Karunasena & Rameezdeen, 2010). The impact of undue selection not only is limited to the settled community but also affects the host community. Brun (2009) added that host communities may become impoverished if their location and resources are shared with newcomers without adequate support or compensation. Although it is difficult to satisfy all essential requirements in selecting a resettlement's location, one needs to take into consideration matters of livelihood and lifestyle of both host and displaced communities to a considerable extent to facilitate their adaptation to the new environment. Table 9.1 synthesizes the long-term adaptability issues identified in the literature.

Table 9.1 Main factors affecting the adaptation of resettled communities.

Main factors	Subfactors	Authors
Housing	Local-climate-adaptable houses	Barenstein (2015)
	Incompatible house design (functionally, socially, and culturally inappropriate)	Ahmed (2011), Andrew et al. (2013), Barenstein (2015), Chang, Wilkinson, Potangaroa, & Seville (2011), Gunawardena & Wickramasinghe (2009), Karunasena & Rameezdeen (2010)
	Quality of houses (durability, space availability)	Ahmed (2011), Chang et al. (2011), Karunasena & Rameezdeen (2010)
	Communal-space availability	Andrew et al. (2013)
	Inability to maintain, expand, and upgrade the structure	Orencio & Fujii (2013)
Infrastructure	Inadequate sanitation	Badri, Asgary, Eftekhari, & Levy (2006)
	Access to physical infrastructure (drinking water, electricity, roads, common buildings, schools, etc.)	Andrew et al. (2013), Gunawardena & Wickramasinghe (2009), Laugé, Hernantes, & Sarriegi (2015), Thalayasingam (2009)
	Reduced availability of community resources (medical, educational, etc.)	Cao, Hwang, & Xi (2012), Cernea (1995), Foresight (2011), Magis (2010), Manatunge, Herath, Takesada, & Miyata (2009), Muggah (2000)
	Lack of transportation network	Cutter et al. (2008), Kusumastuti, Viverita, Husodo, Suardi, & Danarsari (2014)
Location	Resettlement in unfamiliar and inhospitable locations	Andrew et al. (2013), Robinson (2003)
	Vulnerability to environmental changes	Foresight (2011)
	Changes in land-use patterns	Ruiz & Vargas-Silva (2013)
	Distance from the previous location/livelihood	Gunawardena & Wickramasinghe (2009), Jha, Barenstein, Phelps, Pittet, & Sena (2010), Lakshman & Amirthalingam (2009), Manatunge et al. (2009)
	Land-ownership/title issues	Barenstein (2015), Godamunne (2012), Gunawardena & Wickramasinghe (2009), Koria (2009), Orencio & Fujii (2013)

The literature explains some of the reasons why people end up being dissatisfied with resettlements. One of these is that the potential of the built environment to ensure effective social mixing is not recognized in the process of resettlement. The possibilities of creating an adaptable built environment along resettlement processes need to be comprehensively addressed to provide a durable solution that satisfies the settlers and sustains their new setting. Based on case studies in Sri Lanka, this research focused on DIDR as a contribution to the knowledge pool as well as to the practice.

5. The empirical study in Sri Lanka

While being a system itself, the built environment is a subsystem of a community. As discussed earlier, the constructs of the built environment—location, housing, and infrastructure—and the interactions of the community continuously shape each other (Lawrence & Low, 1990). To understand the nature of this two-way relation, it is essential to know how people presume the reality. However, there can be an unknown reality that is not directly observable. Having assumed that, this research took critical reality as its philosophical position. Critical realists believe that following an experience through the sensations that shape the observable events (i.e., the underlying reality) can be mentally processed by reasoning backward (Saunders, Lewis, & Thornhill, 2016). To appreciate the complex nature of postdisaster resettlements and community satisfaction, we explored it from the standpoint of the built environment. Our main assumption was that resettlement dissatisfaction could be interpreted through observable factual events. Thus, we followed a retroductive approach to identify the latent mechanism of resettlement dissatisfaction by working back from observed regularities in the built environment to formulate a possible explanation.

In this research, we adopted the multiple embedded case-study design (Yin, 2014), seeking to confirm the following assumption: Resettlement dissatisfaction is related to the desire to possess a house. We infer that the dissatisfaction on resettlement processes follows the same pattern in both conflict-induced and environment-induced displacements and resettlements. To validate this supposition, we selected cases representing each of the two conditions. Accordingly, the cases comprise groups of people who lived or were expecting to live in permanent disaster-induced resettlement schemes in Sri Lanka. We selected three districts as study areas: Jaffna, a conflict-stricken district in the Northern province; Batticaloa, one of the districts most affected by the 2004 tsunami in the Eastern province; and Badulla, a district highly prone to landslides in the Uva province (Fig. 9.2).

We approached the local authorities in each district and selected the initial cases based on their recommendations. We further expanded the number of cases using a snowball-sampling technique, through the recruitment of additional contacts from among acquaintances of early respondents (see Table 9.2 for a characterization of the studied cases). We conducted semistructured interviews with affected people, their hosts, and disaster experts in the selected communities. Data saturation and unanimity of the respondents determined the number of respondents. All the ethical procedures agreed with the University of Huddersfield's ethics committee were followed during the data collection.

We then employed template analysis, a type of thematic analysis aligned with the epistemological position of the research, to reach our conclusions. The key analytical themes comprised the favorable and unfavorable conditions of the resettlements, the procedures adopted during the resettlement process, and the expectations and requirements of the beneficiaries in their new settlement. After the detailed analysis of the nine cases, we explored some possible theoretical generalizations and identified common variables across cases to develop the arguments presented in Section 7.

Figure 9.2 The location of the three case-study areas. The authors.

Table 9.2 Profile of the studied cases.

District	Type of disaster (date)	Postdisaster resettlement site	Resettlement phase	Integrated into a host community	Voluntariness of the resettlement	Number interview
Batticaloa	Tsunami (2004)	Kallady	Resettled (more than ten years ago)	Yes	Involuntary	10
		Thiraimadu	Resettled (more than ten years ago)	No	Involuntary	13
		Kaluwanchikudy	Resettled (more than ten years ago)	Yes	Involuntary	15
Badulla	Landslide (2006–2011)	Arnhall	Expecting resettlement	N/A	Involuntary	6
		Queenstown	Expecting resettlement	N/A	Involuntary	5
		Makaldeniya	Houses allocated	No	Involuntary	7
		Meeriyabedda	Expecting resettlement	N/A	Involuntary	4
		Newburgh, Ella	Houses allocated	No	Involuntary	7
Jaffna	Civil war (1983–2009)	New Moor Street	Resettled (one year ago)	Yes	Voluntary	6
Total						73

6. The DIDR context of Sri Lanka

With a national disaster toll comprising about 35,000 deaths and 70,000 destroyed and 40,000 damaged houses (World Bank, 2014), the Indian Ocean tsunami of December 26, 2004 jump-started large-scale DIDR in Sri Lanka. The disaster eventually triggered the displacement of more than 500,000 people (Jayasuriya & McCawley, 2010). According to Barenstein and Leemann (2012), before this event, Sri Lanka lacked a well-formulated housing policy for DIDR. Governmental institutions thus dealt with the tsunami resettlements on an ad hoc basis, following policies guided by situational requirements. Under these circumstances, the Sri Lankan government faced significant time and capacity pressures to address the DIDR challenges. Due to these constraints, most of the DIDR decisions and procedures were stand-alone processes—some of which were successful while others failed to achieve the intended outcomes.

A well-known example of the latter is the implementation of the "buffer-zone policy." This policy restricted the access to the lands lying within 300—500 m from the sea (depending on the coastal conditions), and hence made many people landless and homeless. This turned housing an ever pressing need. The sense of urgency at this stage pushed governmental authorities to implement resettlement interventions as early as possible and led to the approval of national and international NGOs' related assistance and funding without due monitoring. Nevertheless, the government could not house all the affected people as planned, due to issues such as the scarcity of suitable land and the potential negative impacts to the livelihoods of some resettled communities (e.g., fisherfolk). Hence, the adopted alternative was to reduce the width of the buffer zone to house the resettlers again in overpopulated coastal regions. This violated the very purpose of the buffer-zone policy, as the potential damage from a similar disaster is generally greater in areas with a high population density.

The lack of experience in Sri Lanka in dealing with large-scale disasters before the 2004 tsunami also contributed to the absence of planned resettlements. The required institutional capacities and arrangements were either inexistent or unprepared to effectively tackle such demands. After the 2004 tsunami, the civil war ravaging the northern part of the country since 1983 aggravated, leading to further displacements. Although this conflict ended in 2009, its housing consequences lasted longer. Besides, the occurrence of devastating landslides and floods at varying intervals worsened the housing deficit. By 2010, resettlements had become a mainstream concern in the national administrative circles, and ministries, legislation, policies, and guidelines were put in place to address DIDR in a more organized manner. To understand the lack of prior experience in dealing with DIDR in Sri Lanka, one needs to grasp the overlapping structure and evolution of governmental institutions in charge of housing, infrastructure, and disaster-related issues.

The main organization responsible for managing the impacts of disasters, the National Council for Disaster Management, was only established in 2005 in response to the 2004 tsunami. Yet, the Resettlement Authority, which would be in charge of postdisaster resettlement processes, was constituted only in 2007. Thus, the initial resettlement activities in the aftermath of the 2004 tsunami were carried out by the ministry in charge of social services. Prior to the establishment of the Resettlement Authority, the National Housing Development Authority, set up in 1979, had some vaguely defined powers and functions regarding the resettlement of displaced persons. This organization prepared and executed proposals, plans, and projects attending to persons already displaced or likely to be displaced through its interventions. Nonetheless, other older governmental organizations—e.g., the National Housing Department (established in 1952), the National Housing Development Department (established in 1977), the Greater Colombo Economic Commission (established in 1977), and the Urban Development Authority (established in 1978)— also have a role in formulating housing policies and carrying out the related projects. Since these policies were not specifically designed to deal with issues linked to a large-scale disaster such as the 2004 tsunami, the authorities struggled with mapping the entities and policies of the newly established organizations that become in charge of older provisions. Understandably, policy-makers and implementers had to take time to resolve these issues.

The more recent National Policy on Disaster Management (established in 2010) stipulates that the needs of disaster-affected people have to be addressed according to national and international guidelines. However, the lack of collaboration and power sharing between many statutory bodies, as well as the surfeit of guidelines and procedures, has been hindering the implementation of effective and efficient resettlements. Practically, it has been a challenge to decide upon and follow the most appropriate guidelines and procedures due to the complexities of the different disasters and landscapes involved. Consequently, the various resettlement-implementation agencies have been randomly adapting existing policies and guidelines on an ad hoc basis, compromising the quality and longevity of these interventions.

In addition, civil-society and private-sector organizations have also been involved in risk-reduction and postdisaster-recovery processes, but they must obtain the consent of the relevant ministry before engaging in such activities. Again, as the procedures and guidelines that they need to follow are often unclear, the level of policy implementation has been poor. This, nonetheless, does not suggest that a rigid, highly centralized, and fully structured disaster, housing, and development policy would have worked effectively in an extreme situation like the 2004 tsunami. Indeed, in such a circumstance, there is a good chance that even the most prepared countries would unsuccessfully respond due to the scope of the issues to face. In Sri Lanka, the major challenge was related not only to the introduction of ad hoc policies but also to the lack of preparedness and expertise in view of the unexpected level of the devastation. Currently, the relevant authorities are attempting to circumvent the vagueness and ambiguities in the DIDR procedures by amending them as appropriate. However, there is still significant room for improvements.

7. The underlying mechanism of resettlement dissatisfaction

The inefficiencies in the DIDR system, as discussed earlier, undoubtedly do not contribute to meeting the expectations of the resettled communities. However, the reasons for such dissatisfaction need to be explored in greater detail so as to support a comprehensive solution to the issues at hand. Resettlement processes in Sri Lanka mainly follow a top-down, centralized approach to planning and implementation, with only very limited inputs from ground-level stakeholders. At the other extreme, bottom-up, decentralized approaches imply the execution of resettlement programs with the participation of communities and local-level organizations. Both approaches have advantages and barriers, leading to different levels of beneficiaries' satisfaction.

To explore this issue, we mapped the observed level of desire to possess a house of disaster-affected people across the postdisaster phases as a proxy to resettlement satisfaction. In this study, the level of *the desire to possess a house* is an indicative qualitative measure and cannot be calculated as a numeric parameter. With its mapping, we did not intend to represent the desire to possess a house as a precise scale but to illustrate its volatility across the various resettlement phases. Fig. 9.3 illustrates the relationship between the resettlement phases and the observed levels of desire to get a house.

We captured the desire level through the answers provided by the resettler participants during the qualitative interviews. We asked them to reflect on questions such as: "What were the reasons behind your decision to move into the new house?"; "why did you decide the leave your new house?"; and "why did you move back to your original house?" Their replies provided the

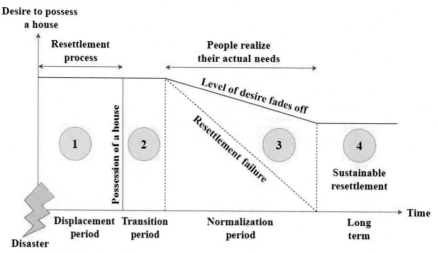

Figure 9.3 The underlying mechanism of resettlement dissatisfaction. The authors.

basis for us to build an explanatory framework to understand the underlying reasons for resettlement dissatisfaction in Sri Lanka across the various stages of resettlement processes.

7.1 Resettlement phases and people's expectations

The resettlement process is procrastinated for people who continue to live in their original homes after a disaster, as the housing urgency is not appreciated by the authorities or the affected communities themselves. In this circumstance, the postdisaster resettlements were more organized and procedural compared with resettling communities who were living in temporary shelters. The reason for the difference is that, following a disaster, the affected people lack alternatives and live in temporary shelters, thus imposing urgency upon the authorities. Also, disaster events attract funds from local and international charity organizations to invest in housing reconstruction. Our model hence adopts the disaster event as the start of the resettlement process.

Phase 1 of the resettlement process comprises the displacement period. At this stage, some lands became unusable or unsafe for habitation and were consequently declared as prohibited zones. Thus, the affected people started to live in temporary shelters until the competent authorities provide a resettlement program. These authorities recognized the need for resettlement during this period. Furthermore, the officials felt a sense of urgency to provide resettlement programs to the affected people as the government intervention is a factor in revoking their choice to return to their old houses.

The desire to get a house is higher during the displacement period—the disaster survivors in temporary shelters expressed the need for a permanent house more than disaster survivors who continued to live in their original houses. The restrictions to build or rebuild permanent structures in disaster-stricken locations impeded the affected people to renovate their fully or partially damaged houses. Since disasters affect most heavily deprived groups, multiple house ownership was uncommon. Before the disaster, most of the disaster survivors lived in temporary, semipermanent, or line houses, lacking afterward immediate alternative accommodation. Thus, they depended on the government for resettlement, and the possession of a permanent house became a priority.

The fear of experiencing another disaster is another reason for the high desire to get a house during the displacement period. People who have suffered a disaster experience panic and depression, as they fear similar future events more than those who have not. This situation validates the priority given to provide resettlement in a safe location for disaster survivors. Their expressed expectation for a permanent house and willingness for a resettlement were high, owing to the traumatic disaster experiences in their previous places.

The pleasure of getting an asset for free also feeds the high desire to get a house during the initial period. As most of the affected people were incapable of acquiring land and houses on their own, they saw in the disaster and the consequent restrictions an opportunity to own a house for free. The officials spotted some forgeries when the number of families claiming for houses exceeded the number of houses being built. There was also evidence of unaffected families living in the temporary shelters for the sake of getting a house. Accordingly, the postdisaster needs for urgent housing might have been artificially inflated to a certain level, although they remained a primary concern.

The interviewees' accounts referred to what they considered to be favorable conditions in the displacement period. In the Kallady case, they particularly mentioned the possession of a land with a permanent house. In R7's words, *"we did not own a house previously; we were living in a rented house before the tsunami. It feels good to own a land with a house."* Agreeing with this statement, R4 stated that *"we lived in a small hut along the coast. It was just an enclosed space with a roof, a cardboard walled the rear of the house. Compared to that, this house is much better."* Indeed, the majority of the resettlers now in this place were once illegal squatters who lived on the coast of Kallady in small semipermanent houses. Therefore, they considered having a land with a permanent house a major asset in their lives. R5 endorsed this by stating: *"the previous land was not mine, I was there unlawfully. Now this place belongs to me. I maintain the land and the house in a way I like."* Although most of them have not yet received their deeds of ownership, they feel a sense of ownership in their new built environment.

The second phase of resettlement comprises the transition period. The beneficiaries were willing to move to the new location as soon as they received the houses. However, it took considerable time to recognize and adjust to the new built environment. During this period, the beneficiaries did not identify their actual needs related to the resettlement. The desire to possess a house remained high during this phase owing to the newness effect. The communities who received the houses more recently, unlike those who resettled a long time ago, assigned an additional value to the houses. We denominated this incremental value as the "newness bias."

During the third phase of the resettlement—the normalization period—the problems and unfavorable conditions identified by the recent settlers differed from those mentioned by long-term settlers. For new settlers, the most unfavorable conditions in this phase was related to the struggle to adapt to a new environment, as they were alien to the surrounding and the associated way of life. On the other hand, some long-term settlers emphasized the wish to own a house due to their intention to sell or rent the property. Indeed, we found abandoned houses mainly in long-term resettlements. In addition, the lack of housing alternatives is another factor that accentuates resettlement needs after a disaster. However, without an increase in income, some beneficiaries found it expensive to maintain or retain the resettlement property in the long term. Subsequently, they sought cheaper options. On the other hand, the restricted zone policies lost their rigor as time passed, making the former locations available for occupation by the beneficiaries. This motivates people to reoccupy their former properties. Thus, the availability of an alternative weakens the will to hold the resettlement property.

7.2 Resettlement procedures and people's expectations

Gaps and barriers in the implementation process also contributed to the decline in the desire to get a house. The adopted top-down procedures presented important shortcomings in several domains, from land and beneficiary selection to house design and construction. These lapses lead to dissatisfaction during the normalization phase, when the beneficiaries realized that their expectations were not fulfilled and they faced adaptation difficulties, losing motivation to hold on to the resettlement property. Long-term settlers expressed preference to return to their former locations or retain the ownership of their former properties. This suggests that the traumatic impression of the survivors about their previous places based on the disaster was a temporary mental state that faded with time as risk salience and risk perception diminished. The additional value assigned by the beneficiaries to the new property declined likewise over time.

Also due to the top-down organizational structure of the resettlement processes, their outcomes often include unfavorable conditions for the beneficiaries, such as inadequate space to extend their houses, defects in the constructions (due to the use of poor-quality materials and inadequate workmanship), and lack of transportation facilities. These conditions eventually hampered the resettlers' adaptability significantly. The beneficiaries also noted that built-environment-related factors (such as the space and layout of the houses) and other unfavorable socioeconomic conditions (such as the resettlement's impact on their livelihoods) could mostly derive from a misalignment between the top-down procedures and grassroots perspectives. Eventually, the combination of adverse conditions and unfulfilled expectations leads to a decline in the desire to hold the property.

The interviewees expressed their disagreement with the adopted resettlement procedures. For instance, the exposure of the resettlement location to flood was mentioned as an unfavorable condition in the Kaluwanchikudy case. According to R26, *"our lands get flooded during rainy seasons,"* about which R25 agreed by saying: *"It usually floods in this area almost every year."* R27 added that *"there are no paths for the water to drain as no drainage facilities were provided. As a mitigation measure, we elevated one side of our land by filling it with earth."* These statements point to the lack of anticipatory measures to address the problem of flooding, although the location was already known to be flood prone. This is among the drawbacks of the current land-selection procedure.

According to the respondents of the Makaldeniya case, the resettlement process was not people centered. They complained that the information given to them about the houses to be built was rather abstract. As put by R45, *"a house plan was shown to us. But we were not given any options, and they did not seek our opinions on the design."* R48 added that *"the plan was shown on a laptop screen. Only four or five people could see it properly. I did not get an idea of the design of the house based on that."* Such statements suggest that although there were some attempts to involve people in the resettlement process, the participatory process was rather hasty and thus ineffective. Also, the beneficiaries were not consulted on the suitability of the houses.

7.3 Resettlement failure as people's expectations change

Despite receiving a new house, the beneficiaries often did not entirely give up the ownership of their former locations. Even beneficiaries who did not own a permanent house before the disaster desired access to their old residence. This implies that they viewed the resettlement property as an additional asset acquired free of charge as a compensation. Also, maintaining the resettlement property was too expensive for some beneficiaries. Thus, the beneficiaries for whom the resettlement property was an additional but costly asset to retain were motivated to make a profit out of it by selling or renting it, thereby diminishing their desire to hold on to the property.

Conversely, some favorable conditions in the resettlement location stimulate the desire to remain in it. For some resettled communities, their previous settlement had worst infrastructure facilities than their new living setting. Hence, the provision of these facilities upgraded their living standard and motivated them to remain in the resettlement. Furthermore, relatively better houses, safe location, and proximity to the previous location and town center fared among the features frequently mentioned as attractions favoring the desire to keep the new property. The final phase of the resettlement comprises sustained long-term resettlement.

The resettlement phase is one of the parameters that influence the positive and negative expectations of the beneficiaries. Thus, it is reasonable to assume that the desire to possess a house may also vary according to changes in the expectations of the beneficiaries. Resettlement failure can occur during the different stages, depending on the changes in the desire of the beneficiaries. Furthermore, the sense of urgency is higher for the disaster survivors living in temporary shelters than for people who continue to live in their own (maybe partially damaged) original houses even if these were located in disaster-prone zones. This shows that soon after a disaster, the affected people see housing as a primary need, all of whom demonstrating willingness to resettle. Thus, the likely decline in this desire is another parameter to explain resettlement failure.

8. Concluding remarks

According to the identified underlying mechanism, in a postdisaster resettlement scenario, the affected persons or families' desire to possess a house is at its peak during the displacement and transition periods. The bases for this desire change over time along with the different phases of the disaster-response activities. The resettlers often realize their actual needs for resettlement at the last phase, when the prevailing reasons to relocate that dominated the first two phases fade off. The decline in the desire to remain in the new house leads to dissatisfactions. Favorable conditions or attracting factors in the resettlement location obstruct this drop, up to a certain extent. However, addressing the causes of the decline is essential to sustain and achieve the primary purpose of the resettlements.

In this sense, to enhance postdisaster recovery and reconstruction and ultimately contribute to the Sendai Framework's Priority 3, resettlements should depart from being viewed as a process of simply providing lands and houses for disaster survivors. The process should not end with the handing over of the houses to the beneficiaries, and governmental officers should embrace the responsibility to monitor the long-term performance of the facilities invested in. This would help to capitalize on the opportunities to learn from experiences and, consequently, to avoid the reproduction of resettlement plans doomed to fail.

References

Ahmed, I. (2011). An overview of post-disaster permanent housing reconstruction in developing countries. *International Journal of Disaster Resilience in the Built Environment, 2*(2), 148–164. https://doi.org/10.1108/17595901111149141

Andrew, S. A., Arlikatti, S., Long, L. C., & Kendra, J. M. (2013). The effect of housing assistance arrangements on household recovery: An empirical test of donor-assisted and owner-driven approaches. *Journal of Housing and the Built Environment, 28*(1), 17–34. https://doi.org/10.1007/s10901-012-9266-9

Badri, S. A., Asgary, A., Eftekhari, A. R., & Levy, J. (2006). Post-disaster resettlement, development and change: A case study of the 1990 Manjil earthquake in Iran. *Disasters, 30*(4), 451–468. https://doi.org/10.1111/j.0361-3666.2006.00332.x

Barenstein, J. D. (2015). Continuity and change in housing and settlement patterns in post-earthquake Gujarat, India. *International Journal of Disaster Resilience in the Built Environment, 6*(2), 140–155. https://doi.org/10.1108/IJDRBE-01-2014-0009

Barenstein, J. D., & Leemann, E. (2012). *Post-disaster reconstruction and change: Communities' perspectives*. Boca Raton, FL: CRC Press.

Berke, P. R., Kartez, J., & Wenger, D. (1993). Recovery after disaster: Achieving sustainable development, mitigation and equity. *Disasters, 17*(2), 93–109. https://doi.org/10.1111/j.1467-7717.1993.tb01137.x

Betts, A. (2009). *Forced migration and global politics*. Chichester: Wiley.

Brun, C. (2009). IDPs and hosts as constitutive categories in protracted displacement: Experiences from Puttalam. In P. Fernando, K. Fernando, & M. Kumarasiri (Eds.), *Forced to move: Involuntary displacement and resettlement—Policy and practice* (pp. 125–144). Colombo: Centre for Poverty Analysis.

Cannon, T. (2008). Vulnerability, "innocent" disasters and the imperative of cultural understanding. *Disaster Prevention and Management, 17*(3), 350–357. https://doi.org/10.1108/09653560810887275

Cao, Y., Hwang, S.-S., & Xi, J. (2012). Project-induced displacement, secondary stressors, and health. *Social Science & Medicine, 74*(7), 1130–1138. https://doi.org/10.1016/j.socscimed.2011.12.034

Cernea, M. M. (1995). Understanding and preventing impoverishment from displacement: Reflections on the state of knowledge. *Journal of Refugee Studies, 8*(3), 245–264. https://doi.org/10.1093/jrs/8.3.245

Chang, Y., Wilkinson, S., Potangaroa, R., & Seville, E. (2011). Donor-driven resource procurement for post-disaster reconstruction: Constraints and actions. *Habitat International, 35*(2), 199–205. https://doi.org/10.1016/j.habitatint.2010.08.003

Combs, D. L., Quenemoen, L. E., Parrish, R. G., & Davis, J. H. (1999). Assessing disaster-attributed mortality: Development and application of a definition and classification matrix. *International Journal of Epidemiology, 28*(6), 1124–1129. https://doi.org/10.1093/ije/28.6.1124

Cronstedt, M. (2002). Prevention, preparedness, response, recovery—an outdated concept? *Australian Journal of Emergency Management, 17*(2), 10–13.

Cutter, S. L., Barnes, L., Berry, M., Burton, C., Evans, E., Tate, E., & Webb, J. (2008). A place-based model for understanding community resilience to natural disasters. *Global Environmental Change, 18*, 598–606. https://doi.org/10.1016/j.gloenvcha.2008.07.013

Dikmen, N., & Elias-Ozkan, S. T. (2016). Housing after disaster: A post occupancy evaluation of a reconstruction project. *International Journal of Disaster Risk Reduction, 19*, 167–178. https://doi.org/10.1016/j.ijdrr.2016.08.020

Foresight. (2011). *Migration and global environmental change*. Final Project Report. Retrieved from http://www.gov.uk/government/uploads/system/uploads/attachment_data/file/287717/11-1116-migration-and-global-environmental-change.pdf.

Freeman, P. K. (2004). Allocation of post-disaster reconstruction financing to housing. *Building Research & Information, 32*(5), 427–437. https://doi.org/10.1080/0961321042000221016

Ganapati, N. E. (2016). Post-disaster housing reconstruction lessons from the 1999 Marmara earthquake, Turkey. In P. Daly, & R. M. Feener (Eds.), *Rebuilding Asia following natural disasters: Approaches to reconstruction in the Asia-Pacific region* (pp. 141–159). Cambridge: Cambridge University Press.

Gilbert, R. (2001). *Doing more for those made homeless by natural disasters.* Disaster risk management working paper series no. 1. Washington, DC: World Bank.

Godamunne, N. (2012). Development and displacement: The National Involuntary Resettlement Policy (NIRP) in practice. *Sri Lanka Journal of Social Sciences, 35–36*(1–2), 37–50. https://doi.org/10.4038/sljss.v35i1-2.7351

Gong, Y., Zhang, R., Yao, K., Liu, B., & Wang, F. (2020). A livelihood resilience measurement framework for dam-induced displacement and resettlement. *Water, 12*(11). https://doi.org/10.3390/w12113191

Grundy-Warr, C., & Rigg, J. (2016). The reconfiguration of political, economic and cultural landscapes in post-tsunami Thailand. In P. Daly, & R. M. Feener (Eds.), *Rebuilding Asia following natural disasters: Approaches to reconstruction in the Asia-Pacific Region* (pp. 210–235). Cambridge: Cambridge University Press.

Gunawardena, A., & Wickramasinghe, K. (2009). Social and economic impacts of resettlement on Tsunami affected coastal fishery households in Sri Lanka. In P. Fernando, K. Fernando, & M. Kumarasiri (Eds.), *Forced to move: Involuntary displacement and resettlement—Policy and practice* (pp. 83–108). Colombo: Centre for Poverty Analysis.

IFRC (International Federation of Red Cross and Red Crescent Societies). (2013). *Post-disaster shelter: Ten designs.* Retrieved from https://www.sheltercluster.org/resources/documents/post-disaster-shelter-ten-designs.

Jahre, M., Persson, G., Kovács, G., & Spens, K. M. (2007). Humanitarian logistics in disaster relief operations. *International Journal of Physical Distribution & Logistics Management, 37*(2), 99–114. https://doi.org/10.1108/09600030710734820

Jayasuriya, S., & McCawley, P. (2010). *The Asian tsunami: Aid and reconstruction after a disaster.* Cheltenham: Edward Elgar.

Jha, A. K., Barenstein, J. D., Phelps, P. M., Pittet, D., & Sena, S. (2010). *Safer homes, stronger communities: A handbook for reconstructing after natural disasters.* Washington, DC: World Bank.

Jigyasu, R., & Upadhyay, N. (2016). Continuity, adaptation, and change following the 1993 earthquake in Marathwada, India. In P. Daly, & R. M. Feener (Eds.), *Rebuilding Asia following natural disasters: Approaches to reconstruction in the Asia-Pacific region* (pp. 81–107). Cambridge: Cambridge University Press.

Joshi, A., & Nishimura, M. (2016). Impact of disaster relief policies on the cooperation of residents in a post-disaster housing relocation program: A case study of the 2004 Indian Ocean Tsunami. *International Journal of Disaster Risk Reduction, 19*, 258–264. https://doi.org/10.1016/j.ijdrr.2016.08.018

Karunasena, G., & Rameezdeen, R. (2010). Post-disaster housing reconstruction: Comparative study of donor vs owner-driven approaches. *International Journal of Disaster Resilience in the Built Environment, 1*(2), 173–191. https://doi.org/10.1108/17595901011056631

Koria, M. (2009). Managing for innovation in large and complex recovery programmes: Tsunami lessons from Sri Lanka. *International Journal of Project Management, 27*, 123–130. https://doi.org/10.1016/j.ijproman.2008.09.005

Kusumastuti, R. D., Viverita, V., Husodo, Z. A., Suardi, L., & Danarsari, D. N. (2014). Developing a resilience index towards natural disasters in Indonesia. *International Journal of Disaster Risk Reduction, 10*, 327–340. https://doi.org/10.1016/j.ijdrr.2014.10.007

Lakshman, R. W. D., & Amirthalingam, K. (2009). Displacement and livelihoods: A case study from Sri Lanka. In P. Fernando, K. Fernando, & M. Kumarasiri (Eds.), *Forced to move: Involuntary displacement and resettlement—Policy and practice* (pp. 57–82). Colombo: Centre for Poverty Analysis.

Laugé, A., Hernantes, J., & Sarriegi, J. M. (2015). Analysis of disasters impacts and the relevant role of critical infrastructures for crisis management improvement. *International Journal of Disaster Resilience in the Built Environment, 6*, 424–437. https://doi.org/10.1108/IJDRBE-07-2014-0047

Lawrence, D. L., & Low, S. M. (1990). The built environment and spatial form. *Annual Review of Anthropology, 19*, 453–505.

Magis, K. (2010). Community resilience: An indicator of social sustainability. *Society and Natural Resources, 23*(5), 401–416. https://doi.org/10.1080/08941920903305674

Manatunge, J., Herath, L., Takesada, N., & Miyata, S. (2009). Livelihood rebuilding of dam-affected communities: Case studies from Sri Lanka and Indonesia. *International Journal of Water Resources Development, 25*(3), 479–489. https://doi.org/10.1080/07900620902957928

Moteff, J., & Parfomak, P. (2004). *Critical infrastructure and key assets: Definition and identification*. Congressional Research Service Report RL32631. Retrieved from http://www.fas.org/sgp/crs/RL32631.pdf.

Muggah, R. (2000). Conflict-induced displacement and involuntary resettlement in Colombia: Putting Cernea's IRLR model to the test. *Disasters, 24*, 198–216. https://doi.org/10.1111/1467-7717.00142

Muggah, R. (2008). *Relocation failures in Sri Lanka: A short history of internal displacement and resettlement*. London: Zed Books.

Oliver-Smith, A. (1991). Successes and failures in post-disaster resettlement. *Disasters, 15*(1), 12–23. https://doi.org/10.1111/j.1467-7717.1991.tb00423.x

Oloruntoba, R., Sridharan, R., & Davison, G. (2018). A proposed framework of key activities and processes in the preparedness and recovery phases of disaster management. *Disasters, 42*(3), 541–570. https://doi.org/10.1111/disa.12268

Orencio, P. M., & Fujii, M. (2013). A localized disaster-resilience index to assess coastal communities based on an analytic hierarchy process (AHP). *International Journal of Disaster Risk Reduction, 3*, 62–75. https://doi.org/10.1016/j.ijdrr.2012.11.006

Quarantelli, E. L. (1995). Patterns of sheltering and housing in US disasters. *Disaster Prevention and Management, 4*(3), 43–53. https://doi.org/10.1108/09653569510088069

Robinson, W. C. (2003). *Risks and rights: The causes, consequences, and challenges of development-induced displacement*. Washington, DC: Brookings Institution.

Ruiz, I., & Vargas-Silva, C. (2013). The economics of forced migration. *Journal of Development Studies, 49*, 772–784. https://doi.org/10.1080/00220388.2013.777707

Saunders, M., Lewis, P., & Thornhill, A. (2016). *Research methods for business students* (7th ed.). Harlow: Pearson Education.

Seneviratne, K., Amaratunga, D., & Haigh, R. (2016). Managing housing needs in post-conflict housing reconstruction in Sri Lanka: Gaps verses recommendations. *International Journal of Strategic Property Management,* *20*(1), 88–100. https://doi.org/10.3846/1648715X.2015.1101625

Staupe-Delgado, R. (2020). Can community resettlement be considered a resilient move? Insights from a slow-onset disaster in the Colombian Andes. *The Journal of Development Studies, 56*(5), 1017–1029. https://doi.org/10.1080/00220388.2019.1626836

Thalayasingam, P. (2009). Conflict, vulnerability and long-term displacement: The case of Puttalam. In P. Fernando, K. Fernando, & M. Kumarasiri (Eds.), *Forced to move: Involuntary displacement and resettlement—Policy and practice* (pp. 111–124). Colombo: Centre for Poverty Analysis.

UNISDR (United Nations Office for Disaster Risk Reduction). (2009). *Terminology on disaster risk reduction.* Geneva: UNISDR.

UNISDR (United Nations Office for Disaster Risk Reduction). (2015). *Sendai Framework for Disaster Risk Reduction 2015–2030.* Retrieved from https://www.undrr.org/publication/sendai-framework-disaster-risk-reduction-2015-2030.

Vidal López, R. C.(Coord.) (2011). *The effects of internal displacement on host communities: A case study of Suba and Ciudad Bolívar localities in Bogotá, Colombia.* Bogotá: Brookings Institution. Bogotá: Brookings Institution – London School of Economics Project on Internal Displacement.

World Bank. (2014). *Lessons learned from the Sri Lanka's tsunami reconstruction.* Retrieved from https://www.worldbank.org/en/news/feature/2014/12/23/lessons-learned-sri-lanka-tsunami-reconstruction.

Yin, R. K. (2014). *Case study research: Design and methods* (5th ed.). Los Angeles, CA: Sage.

10

The media coverage of climate change in Portugal

Alexandre Oliveira Tavares[1], Neide Portela Areia[2], José Manuel Mendes[3] and Hugo Pinto[3]

[1]*Centre for Social Studies and Faculty of Sciences and Technology, University of Coimbra, Coimbra, Portugal;* [2]*Centre for Social Studies, University of Coimbra, Coimbra, Portugal;* [3]*Centre for Social Studies and Faculty of Economics, University of Coimbra, Coimbra, Portugal*

1. The media communication of climate change

1.1 Climate emergency and the social inertia regarding climate change

Due to climate change, the frequency, intensity, and/or duration of climate-related disasters, such as floods, droughts, storms, and extreme temperatures, are undoubtedly increasing (Booth et al., 2020), posing one of the most challenging threats to the world and its communities (Al-Amin, Nagy, Masud, Filho, & Doberstein, 2019). In 2015, the United Nations Office for Disaster Risk Reduction (UNISDR, now UNDRR) outlined the urgency of "addressing climate change as one of the drivers of disaster risk" within its Sendai Framework for Disaster Risk Reduction 2015–2030 (UNISDR, 2015, p. 6). Five years later, the scientific community produced a formal declaration in which more than 11,000 scientist signatories from around the world warned humanity that "clearly and unequivocally (...) planet Earth is facing a climate emergency" (Ripple, Wolf, Newsome, Barnard, & Moomaw, 2020, p. 8). Aligned with science, an exponential rise of "climate emergency" declarations was observed at a political level, including by more than 1000 local governments in 25 countries. Theses initiatives may trigger a new paradigm for climate governance (Davidson, 2020). However, only relying on governments and scientific institutions to cope with climate change may lead to a failure of these institutions' efforts to do so (Yu, Lin, Kao, Chao, & Yu, 2019). Scientists and policy-makers come

Investing in Disaster Risk Reduction for Resilience. https://doi.org/10.1016/B978-0-12-818639-8.00010-7

to recognize that climate change urges action in several societal spheres, including from citizens. Yet, it seems challenging to enhance the societal transition to environmental sustainability, as laypersons' mindsets and behaviors are often considered the most difficult to transform (O'Brien, 2018).

Social inertia regarding the current environmental crisis generally prevails in our societies. Indeed, there is still insufficient public mobilization and engagement to climate change, with individuals not actively adopting proenvironmental behaviors or engaging in climate-related disaster risk reduction, by taking actions to reduce risks in their everyday lives (Brulle & Norgaard, 2019; González-Riancho, Gerkensmeier, & Ratter, 2017).

1.2 The role of the media in engaging civil society in climate change

The critical role and value of scientific information and communication for successful adaptation to climate change and disaster risk reduction and resilience have been recognized on a global level, such as by the Intergovernmental Panel on Climate Change (IPCC) (Corner, Shaw, & Clarke, 2018) and the signatory nations of the Sendai Framework (UNISDR, 2015). However, and despite the international recommendations, research shows that individuals demonstrate a superficial understanding of climate change; their personal concern is low and only increased by immediate threats, such as climate-related disasters (e.g., forest fires) (Moser & Dilling, 2012). This panorama raises a critical question about the effectiveness of the public's education (Moser & Dilling, 2012). In this aspect, media plays a critical role, as it is the primary source of information for the general public and, therefore, is considered the mediator between the scientific community and lay audiences (e.g., citizens and policy-makers) (Fig. 10.1) (Areia, Intrigliolo, Tavares, Mendes, & Sequeira, 2019).

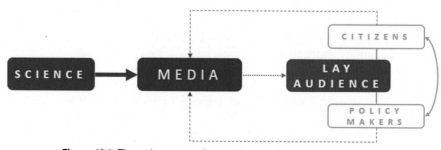

Figure 10.1 The science—media—audience relationship. The authors.

In democratic systems, governments' legislation requires the public accountability and its implementation calls for the active engagement of the public. Therefore, an efficient communication of climate change by the media is an essential link between scientists and laypersons—as science alone cannot compel individuals to action (Moser & Dilling, 2012). For such, a successful reframing of climate change by the media would result in making the complex scientific debate understandable, relevant, and personally important, thus fostering citizens' engagement with the subject (Nisbet, 2009). However, considering the current lack of acknowledgment and engagement of individuals with climate change (Moser, 2010; Moser & Dilling, 2012), it may be inferred that the media has been failing to efficiently reframe the subject and thus properly educate the general public (Smith, 2005). On this matter, Romps and Retzinger (2019) found that the current media's communication praxis lacks the scientific context that the audiences need to make sense of climate-change risk and further actively engage with it.

Indeed, communicating climate change is truly challenging, and its related difficulties may be limiting the crucial role of the media in engaging individuals and policy-makers with the subject. Moser and Dilling (2012) summarized four reasons that may explain the media's unsuccessful communication about climate change: (1) the characteristics of climate change itself; (2) the politicization and institutionalization of climate change; (3) the psychological ways of processing information; and (4) the characteristics of the media's communication.

Concerning this last aspect, framing is frequently pointed out in the literature as a challenging media tool that, when inadequately used, may compromise proper communication (Cissel, 2012; Nisbet, 2009). Media framing is the way in which information is presented to its audiences. According to Entman (1993, p. 51),

> to frame a communicating text or message is to promote certain facets on a 'perceived reality' and make them more salient in such a way that endorses a specific problem definition, causal interpretation, moral evaluation, and/or a treatment recommendation.

Media framing, therefore, has a crucial role in the formation of public opinion and engagement with the issues (Cissel, 2012). As such, it is of utmost importance to understand the media's framing of climate change when considering the urgent need of engaging individuals with it.

For instance, several research works about the role of the media framing on the construction of individuals' opinion on climate change have emerged in the past decade (Boykoff, 2013; Brulle, Carmichael, & Jenkins, 2012; Brüggemann & Engesser, 2017; Hmielowski, Feldman, Myers, Leiserowitz, & Maibach, 2013; Jang, 2013; Nisbet, 2009; Nisbet & Myers, 2007). Most of these research works demonstrated that the media tend to overlook relevant debates, by emphasizing the queries between "warners versus deniers" or the origins' controversies (anthropogenic vs. natural). This form of reporting climate change has contributed to increasing the gap between the scientific accounts and the public's opinion and policy actions. As such, it contributes to maintain the public's confusion and thus leads individuals and policy-makers not to actively follow the mainstream views of the scientific community.

Considering the worrying effects of media framing in the formation of lay audiences' opinion and subsequent engagement with climate change (Boykoff, 2013; Brulle, Carmichael, & Jenkins, 2012; Brüggemann & Engesser, 2017; Hmielowski et al., 2013; Jang, 2013; Nisbet, 2009; Nisbet & Myers, 2007; Spence & Pidgeon, 2010), this study aimed to explore the media coverage on climate change in the Portuguese context between 2017 and 2018. Specifically, it sought to identify the themes most frequently mentioned by the Portuguese press and thus ascertain the possible frames used by the Portuguese journalists. Through this study we intended to identify possible biases in the Portuguese media's discourses about climate change and further discuss its possible influence on the general public and policy-makers' engagement with the subject.

2. Study design

The main goal of this study was to analyze the type of coverage given by the Portuguese online media to the news related to climate change between January 2017 and March 2018. To this end, we applied a quantitative content analysis to a sample of news published in Portuguese media websites, to identify patterns in the country's media communication on climate change. As shown in Fig. 10.2, we carried out the study through four main methodological steps: (1) definition of a research equation; (2) definition of a timeframe; (3) selection of the most relevant news; and (4) quantitative content analyses.

We first defined a research equation, which was then computed into the *Google News* search engine. Then we retrieved the most relevant news articles from Portuguese newspapers. We established four timeframes, considering the period from January

Figure 10.2 Methodology design. The authors.

2017 to March 2018. With this definition, we wanted to ensure the selection of relevant news from different periods of the year.

We selected a maximum of 60 new articles for each timeframe. We conducted the first screening of the news by taking into account their relevance for this study, through preliminary readings of each article: (1) the title, the article's opening paragraph, and first paragraph; or (2) the title and the first two paragraphs, when there was no opening paragraph. The news were eliminated when they were not clearly related to the topic of investigation. Then we conducted an in-depth and detailed analysis of each article, to exclude duplicated articles and to eliminate the ones that did not exactly cover climate-change-related topics.

We conducted a news' content analysis considering two main types of variables: general characteristics and specific characteristics. We considered the news' general characteristics as follows: (1) geographical focus; (2) event's timing; (3) involved actors; and (4) type of knowledge reported. Conversely, we included as the news' specific characteristics: (1) risk-management actions for climate change; (2) adaptation measures for climate change; (3) sector; and (4) hazardous process. For the description of the news articles' contents, the categories' prevalence was determined through frequency statistics. We also used this technique to identify the prevalence of the specificities of actors' discourses. The association between main actors and specific characteristics was calculated through Pearson product—moment correlation (r). All statistical analyses were performed using *IBM SPSS Statistics*, version 22.

3. Climate-change communication: a Portuguese-media portrayal

3.1 General characteristics of climate-change news articles

We selected 217 news articles related to climate change from 34 online newspapers. More than half of the news ($n = 130$, 59.9%) were selected from five newspapers. *Diário de Notícias* ($n = 38$, 17.5%) and *Público* ($n = 35$, 16.1%) covered the majority of the selected news, followed by *Observador* ($n = 25$, 11.5%), *Sapo* ($n = 17$, 7.8%), and *Expresso* ($n = 15$, 6.9%). The remaining 87 news (40.1%) were covered by 29 other newspapers, each gathering 8 or less items. It is worth mentioning that we identified a few environmental or agriculture-specific online newspapers:

one environmental-related (*Tempo.pt*) and one rural-related (*Vida Rural*), covering 1 (0.5%) and 5 (2.3%) items, respectively.

We analyzed the contents of the selected articles considering the general and specific characteristics of each news article. Table 10.1 displays the news' general characteristics, regarding the geographical focus, event's timing, actors, and type of knowledge.

Regarding the **geographical focus**, national events (e.g., the political debate on the Free Eucalyptus Law and its consequences for the forest fires, or the low water levels of Portuguese dams) were the topics most covered by the Portuguese media. More than a quarter of the news did not focus on a specific geographical context, by giving salience to global and general events (e.g., the Paris Climate Accords, or the general agricultural-production fall due to climate change). Concerning the **event's timing**, most of the news cover present events, such as Donald Trump's withdrawal from the Paris Climate Accords or the Portuguese forest fires of June and October 2017.

Taking into account the main **actors** of the Portuguese media coverage, three quarters of the news articles highlighted the discourses of managers/technicians (e.g., the alert from the National Council of the Environment and Sustainable Development to

Table 10.1 News' general characteristics, n (%).

Geographical focus						
National	Europe	North America[a]	Latin America	Africa	Others	Global
109 (50.2%)	13 (6.0%)	16 (7.4%)	7 (3.2%)	11 (5.1%)	5 (2.3%)	56 (25.8%)

Event's timing		
Present	Potential	Predictable
178 (82.0%)	14 (6.5%)	25 (11.5%)

Involved actors			
Managers/technicians	Policy-makers	Researchers	Users/general public
83 (38.2%)	79 (36.4%)	43 (19.8%)	12 (5.5%)

Type of knowledge			
Technical	Political	Academic	Nonexpert
95 (43.8%)	67 (30.9%)	43 (19.8%)	12 (5.5%)

[a]Without Mexico.

communities on the need to save water due to the extreme Portuguese drought) and of policy-makers (e.g., the Government debate on farmers' compensation after forest fires). As a result, technical knowledge and political knowledge are predominant in the Portuguese media coverage. The Portuguese media gives very little salience to users/general public and thus to news based on nonexpert knowledge.

3.2 Specific characteristics of climate-change news articles

We analyzed the specific characteristics considering the following themes: risk-management actions for climate change, adaptation measures for climate change, main sector, and hazardous process. Table 10.2 displays the frequency and prevalence results of each category of these themes.

Table 10.2 News' specific characteristics, n (%).

Risk-management actions for climate change

Impact reduction/mitigation	Alert	Reactive/emergency	Preventive	None
85 (39.2%)	36 (16.6%)	27 (12.4%)	26 (12%)	43 (19.8%)

Adaptation measures for climate change

Prevention of environmental degradation	Access to goods and services	Security of people and belongings	Reduction of economic loss	None
70 (32.3%)	47 (21.7%)	36 (16.6%)	23 (10.6%)	41 (18.9%)

Sector

Agriculture	Forest	Health	Energy	Biodiversity	Coastal areas/sea	Others[a]	Several	None
37 (17.1%)	20 (9.2%)	17 (7.8%)	10 (4.6%)	9 (4.1%)	14 (6.5%)	18 (8.3%)	34 (15.7%)	58 (26.7%)

Hazardous process

Drought	Forest fires	Sea agitation/ coastal threats	Extreme meteorological conditions	Floods	Winds/ tornados	Several	None
75 (34.6%)	23 (10.6%)	10 (4.6%)	10 (4.6%)	1 (0.5%)	1 (0.5%)	28 (12.9%)	69 (31.8%)

[a]Economy ($n = 8$, 3.7%); security ($n = 8$, 3.7%); transports and communication ($n = 2$, 0.9%).

Concerning **risk-management actions for climate change**, impact reduction/mitigation was the most mentioned action. Remarkably, almost 20% of the news did not refer to any risk-management action. The prevention of environmental degradation accounted for the most mentioned **adaptation measures for climate change**, whereas the reduction of economic loss was, surprisingly, barely mentioned. Similar to risk-management actions, almost 20% of the news did not mention any adaptation measure.

More than a quarter of the selected news did not cover any **sector**. When a specific sector was mentioned, agriculture was given more attention. On the other hand, some important sectors such as economy or security were barely mentioned in the Portuguese media coverage. Finally, regarding the main **hazardous process**, drought was the most frequently mentioned, followed by the news that did not cover any hazardous process.

3.3 The actors' discourses conveyed by the Portuguese media

We conducted a detailed analysis of the actors' discourses, to identify the most dominant themes considering risk-management actions, adaptation measures, sector, and hazardous processes. Particularly, we analyzed the discourses of policy-makers, researchers, and managers, taking into account the prevalence and correlations with the categories of each theme. As users are barely mentioned in the Portuguese media coverage, they were excluded from the following analysis. Fig. 10.3 displays the prevalence (%) of each theme's categories in the actors' discourses, and Table 10.3 presents the correlations between actors and theme's categories.

Considering the **policy-makers**' discourses, and regarding **risk-management actions**, we found a positive moderate correlation with the lack of any mention of risk-management actions, $r = 0.35$, $P < .001$, which was the most frequent category found in their discourses, alongside "impact reduction/mitigation actions." Still regarding risk management, we found a negative weak correlation with "alert" actions. Taking into account the **adaptation measures**, we found negative and weak correlations between policy-makers and the "prevention of environmental degradation," and "security." We found a positive moderate correlation with the lack of any mention of adaptation measure, which also showed the highest prevalence ($n = 30$, 38%). Considering the **sector**, we found negative weak correlations between

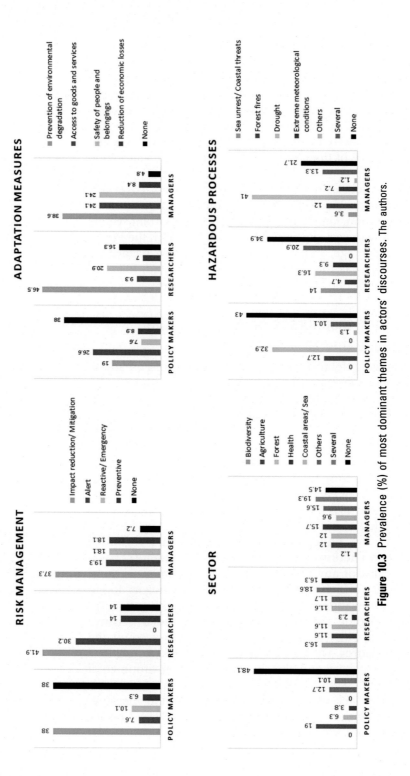

Figure 10.3 Prevalence (%) of most dominant themes in actors' discourses. The authors.

Table 10.3 Pearson correlations between actors and dominant themes in their discourses.

Risk management

	Impact reduction/mitigation	Alert	Reactive/emergency	Preventive	None
Policy-makers	−0.02	−0.19[b]	−0.05	−0.13	0.35[b]
Researchers	0.03	0.18[b]	−0.19[b]	0.03	−0.07
Managers	−0.03	0.06	0.14[a]	0.15[a]	−0.25[b]

Adaptation measures

	Prevention of environmental degradation	Access to goods and services	Security	Reduction of economic loss	None
Policy-makers	−0.22[b]	0.09	−0.18[b]	−0.04	0.37[b]
Researchers	0.15[a]	−0.15[a]	0.06	−0.06	−0.03
Managers	0.11	0.05	0.16[a]	−0.06	−0.28[b]

Sector

	Agriculture	Forest	Health	Energy	Biodiversity	Economy	Security	Transport	Coastal areas/sea	Several	None
Policy-makers	0.04	−0.08	−0.11	−0.03	−0.16[a]	0.004	0.004	0.03	−0.20[b]	−0.12	0.37[b]
Researchers	−0.07	0.04	−0.10	−0.05	0.30[b]	0.03	0.03	−0.05	0.11	0.04	−0.12
Managers	−0.11	0.08	0.23[b]	0.10	−0.12	−0.003	−0.003	0.02	0.10	0.08	−0.22

Hazardous process

	Drought	Forest fires	Sea agitation/coastal threats	Extreme meteorological conditions	Floods	Several	None
Policy-makers	−0.03	0.05	−0.17[a]	−0.17[a]	−0.05	−0.06	0.19[b]
Researchers	−0.19[b]	−0.10	0.22[a]	0.11	−0.03	0.12	0.03
Managers	0.11	0.04	−0.04	0.10	0.09	0.01	−0.17[a]

[a]Correlation is significant at a level of 0.05 (with a two-tailed hypothesis).
[b]Correlation is significant at a level of 0.01 (with a two-tailed hypothesis).

policy-makers and "biodiversity," and "coastal areas/sea," whereas we found a positive moderate correlation between mentioning any sector, $r = 0.37$, $P < .001$. Indeed, almost half of the governmental discourses do not mention any sector in particular. Finally, considering the **hazardous process**, we found negative weak correlations between policy-makers and "sea agitation/coastal threats," and "extreme meteorological conditions." Policy-makers tend not to mention any hazardous process, which was also demonstrated by a positive correlation with this category, $r = 0.19$, $P < .001$.

With regard to **researchers'** discourses, and considering **risk-management actions**, "impact reduction/mitigation" was the most frequent action found in academics' discourses. Also, we found a positive weak correlation between "alert," whereas we found a negative weak correlation between "reactive/emergency." Concerning the **adaptation measures**, the "prevention of environmental degradation" seemed to be a priority in researchers' discourses, also shown by a positive correlation. Still, we found a negative weak correlation between "access to goods and services." Considering the **sector**, researchers tend to mention various sectors simultaneously, but we found a positive moderate correlation between "biodiversity" in particular. Finally, researchers tend not to mention any hazardous process and we found a negative weak correlation between "drought," whereas we found a positive weak correlation between sea agitation/coastal threats.

Lastly, **managers** tend to more frequently mention "impact reduction/mitigation" when considering **risk-management actions**. Also, we found positive correlations between "reactive/emergency" actions, and "preventive" actions. We also found a negative correlation between managers and mentioning any risk-management action. With regard to the **adaptation measures**, "prevention of environmental degradation" was most frequently mentioned by managers. Still, we found a positive correlation between "security," whereas we found a negative correlation when any measure was mentioned, $r = 0.28$, $P < .001$. Considering the **sector**, managers tend to frequently mention "several" sectors simultaneously or the "health" sector in particular, which was also demonstrated by a positive correlation. Finally, drought was the **hazardous process** most commonly mentioned by managers, and we found a negative correlation with mentioning any hazardous process.

4. The stenographic media communication of climate change and its influence on lay audiences

Fig. 10.4 summarizes the main results of our study and the possible influence of the Portuguese media on the general public and the government.

We hardly found independent or specific newspapers (e.g., environment focused) during our data collection. This substantiates the idea that climate-change communication in Portugal is reported by generalist journalists who do not have a sophisticated knowledge about environmental issues. Consequently, they tend to uncritically rely on outside experts and report back this information to the general public (Boykoff & Yulsman, 2013). This panorama may have, a priori, an obvious consequence: biased climate-change communication.

Indeed, similarly to the global trend (Boykoff & Yulsman, 2013), the Portuguese media seems to give superficial and simplistic attention to climate change, which is actually quite a

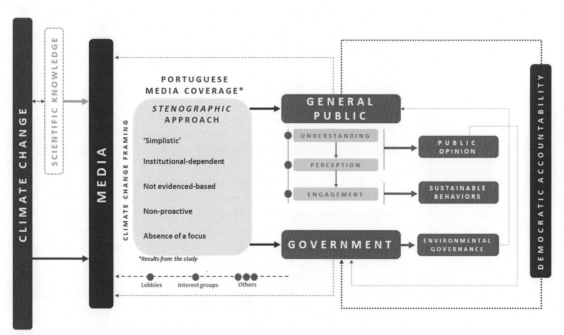

Figure 10.4 The Portuguese media's potential influence on climate-change opinions, behaviors, and governance. The authors.

complex and nuanced subject. Rather than providing an accurate coverage of climate change by conveying scientific-based information, the Portuguese media seems to use a "stenographic approach," as described by Boykoff and Yulsman (2013, p. 367), framed in an "environmental hyperopia" (Uzzell, 2019, p. 314), to communicate climate change (Tavares et al., 2020). Rather than engaging the general public with scientific evidence, the Portuguese media seems more likely to highlight technological/ political debates on climate change, without focusing on clear resilient actions (risk-management and/or adaptation measures) and disclaiming the heterogeneity of the affected sectors and hazardous processes. Instead, it would be preferable to give more salience to the discussion of efficient risk-management/ adaptation measures, grounded on existing international recommendations (e.g., the Sendai Framework), to enhance the public understanding of climate-related disaster risk and to strengthen climate governance at a national (or regional) level (UNISDR, 2015). Concretely, a greater focus of the media on resilient measures would properly educate the general public to cope with forest fires or other climate-change-related hazardous processes, which would in turn contribute to the development of communities' resilience to them. Also, regarding the international focus, a significant space has been given to Donald Trump's withdrawal from the Paris Climate Accords, whereas little attention has been given to the mitigation policies determined in this global agreement. Naturally, this stenographic way of doing journalism—by focusing on a biased technological/political debate—may skew the public's understanding of the evidence-based climate science. Consequently, it may contribute to the perpetuation and even increase of the public's uncertainty about the urgency of climate change, and thus to the disengagement of citizens from the subject (Patt & Weber, 2013).

In fact, framing climate change as an institutionalized debate is not a novelty in the Portuguese press. Previous studies also demonstrated that the Portuguese media tends to give more space to climate-change-related policy-making, followed by technological discussions (Areia et al., 2019; Horta & Carvalho, 2017; Horta, Carvalho, & Schmidt, 2017). In view of these results, it is worth mentioning that the declarations of the main actors in the media—in this case, policy-makers and technicians/ managers—are closely related to the social construction of climate change (Carvalho, 2008).

For such, it was of utmost importance to deepen the knowledge about the discourses of the most relevant social actors covered by the Portuguese media. From a detailed analysis of the main actors' discourses, it was possible to verify that while managers show a concrete but complex discourse—focused on specific measures, sectors, and hazardous processes—the discourses of policy-makers are mainly bare. Additionally, despite the little attention given to the academic discourses in the Portuguese press, it was possible to verify that researchers consider concrete risk-management actions and adaptation measures. Also, researchers tend to give relevance to heterogeneous sectors and hazardous processes, some of them hardly mentioned by managers or policy-makers, such as coastal areas and threats. Therefore, we may infer that if more science-based climate information would be mentioned in the Portuguese press, by researchers' discourses, for instance, individuals would be able to access more reliable and concrete information on climate change through mainstream newspapers. This would help close the well-documented science—action gap (Moser & Dilling, 2012), by increasing citizens' climate-change understanding, perception, and engagement. Subsequently, it would foster individuals' adoption of more sustainable behaviors (e.g., changes in energy-consumption patterns, travel modes, lifestyles, reproductive choices) (Spence & Pidgeon, 2009; Wolf & Moser, 2011).

The Portuguese panorama of stenographic climate-change information reporting, characterized by an unclear or absent focus and by not being evidenced-based, may have two major consequences for both the general public and policy-makers, according to Kahan (2014). On one hand, it frustrates individuals' collective interest in issues related to climate change. On the other hand, governmental institutions may be less likely to adopt public policies that reflect the best available empirical evidence on climate change (Kahan, 2014). Additionally, we argue that these media biases may jeopardize the democratic role of the Portuguese press. Indeed, the current stenographic climate journalism may compromise the relationship between the general public and policy-makers postulated by democracy, by hampering the public's accountability in climate governance.

The public accountability can be considered "the opportunity of citizens to critically monitor and debate proceedings of political decision-making," which implies that policies are scrutinized, discussed, and criticized in public (Karlsson-Vinkhuyzen, Friberg, & Saccenti, 2016; Steffek, 2010). At this point, the media, as a part of democracy, plays a critical role in educating individuals to further participate in policy-making. Indeed, it is expected that, in democracies, the media should at least be a stimulus for individuals to learn, choose, and become actively involved in political processes (Gurevitch & Blumler, 1990). However, considering the biased Portuguese climate-change media coverage, mainly based on bare information and institutional-dependent discourses (disclaiming the role of individuals, for instance), we infer that the Portuguese media may be contributing to maintain citizens "rationally ignorant," as stated by Moravcski (2004, p. 344). Consequently, this form of climate-change reporting does not strengthen the accountability relationship between the political system and the general public, which is desirable from the perspective of democracy.

To sum up our results and discussion, it seems that the consequences of the Portuguese stenographic media coverage of climate change go beyond the individuals' illiteracy about the subject. Indeed, it may not only compromise the public's engagement with climate change and its risks, and adoption of more sustainable behaviors, but also harm the democratic accountability relationship between the general public and governments.

Considering our study's results, we argue that the investment in climate action (European Commission, 2019; UN General Assembly, 2015) and disaster risk reduction for resilience (UNISDR, 2015) should better take into account the communication dimension (Areia, Tavares, & Mendes, 2021), by building the media's capacity to put into practice an environment-specialized journalism. More space would then be given to science-based information and to a wider range of actors' discourses (e.g., general public, researchers). An environment-specialized journalism would also give more salience to proactive and less institutional-dependent coping actions for climate change and show a holistic perspective of environmental issues (e.g., social and economic dimensions).

It is worth mentioning some of the challenges posed to the Portuguese media, which Kahan (2014, p. 11) describes as "the tragedy of the science communication commons." Specifically, communicators must reexamine their assumptions that traditionally framed climate change, in order to reduce the journalistic bias and thus enhance individuals' engagement with the subject and empower their accountability regarding climate governance. However, as stated by Moser and Dilling (2012, p. 169), "communication on climate change is only part of the picture." Despite the central role of the media in raising awareness of individuals, this might not directly result in behavior change or policy actions. Indeed, to ensure an effective and evidence-based communication and thus an active engagement of individuals vis-à-vis climate change, the communication by the media must be supported by policy, economic, and infrastructure changes. This would enable journalists to properly and relevantly report climate information (e.g., greater investment in environment-specialized journalism), which, in turn, may enhance individuals' sustainable actions (e.g., facilitation of access to alternative energies) and individuals/communities' resilience to climate-related risks (e.g., participatory planning and adaptive management to enhance communities' resilience to climate change) (Moser & Dilling, 2012; Ockwell, Whitmarsh, & O'Neill, 2009).

5. Concluding remarks

Through this study we showed that the Portuguese press reports stenographic accounts on climate change, neither conveying a holistic perspective nor encompassing all relevant actors, such as citizens or members of the scientific community. The study also highlighted the existence of a passive attitude toward climate-change impacts, mainly dependent on bare debates. We discussed the consequences of this biased media coverage, especially regarding its relation to individuals' lack of understanding and thus disengagement with climate change, as well as its negative impacts on the democratic-accountability relationship between the public and environmental governance. We argue that to overcome these gaps in the Portuguese journalism landscape, more science-based climate information should be conveyed to the general public. This could help to enhance individuals' engagement with the subject and facilitate embracing more sustainable behaviors, to build individuals/communities' resilience to climate-related risks, and to increase individuals' informed participation in environmental governance.

Our study has several limitations, mainly regarding its methodological design (e.g., short timeframe), sample size, and data analysis. Future studies may consider replicating this research by using a larger sample and conducting in-depth qualitative data analyses. Also, a comparative analysis with other countries would be important to identify differences in climate-change media coverage and its possible effects on public awareness. Future research should also go beyond the mainstream media communication and explore the role of digital social media (e.g., *Twitter*) in engaging the general public in the climate-change debate through, for instance, online discussions. Likewise, deeper analyses should be conducted to better understand the mediation relationship between science and mass media, and the general public and policy-makers.

Acknowledgments

This study was part of the RiskAquaSoil project, cofinanced by the European Regional Development Fund (FEDER) through the Cooperation Program INTER-REG Atlantic Space, with reference EAPA_272/2016.

References

Al-Amin, A. Q., Nagy, G. J., Masud, M. M., Filho, W. L., & Doberstein, B. (2019). Evaluating the impacts of climate disasters and the integration of adaptive flood risk management. *International Journal of Disaster Risk Reduction, 39*, 101241. https://doi.org/10.1016/j.ijdrr.2019.101241

Areia, N. P., Intrigliolo, D., Tavares, A., Mendes, J. M., & Sequeira, M. D. (2019). The role of media between expert and lay knowledge: A study of Iberian media coverage on climate change. *Science of the Total Environment, 682*, 291–300. https://doi.org/10.1016/j.scitotenv.2019.05.191

Areia, N. P., Tavares, A. O., & Mendes, J. M. (2021). Environment actors confronting a post climate-related disaster scenario: A feasibility study of an action-based intervention aiming to promote climate action. *International Journal of Environmental Research and Public Health, 18*(11), 5949. https://doi.org/10.3390/ijerph18115949

Booth, L., Fleming, K., Abad, J., Schueller, L. A., Leone, M., Scolobig, A., & Baills, A. (2020). Simulating synergies between climate change adaptation and disaster risk reduction stakeholders to improve management of transboundary disasters in Europe. *International Journal of Disaster Risk Reduction, 49*, 101668. https://doi.org/10.1016/j.ijdrr.2020.101668

Boykoff, M. T. (2013). Public enemy No. 1? *American Behavioral Scientist, 57*(6), 796–817. https://doi.org/10.1177/0002764213476846

Boykoff, M. T., & Yulsman, T. (2013). Political economy, media, and climate change: Sinews of modern life. *Wiley Interdisciplinary Reviews: Climate Change, 4*(5), 359–371. https://doi.org/10.1002/wcc.233

Brulle, R. J., & Norgaard, K. M. (2019). Avoiding cultural trauma: Climate change and social inertia. *Environmental Politics, 28*(5), 886–908. https://doi.org/10.1080/09644016.2018.1562138

Brulle, R. J., Carmichael, J., & Jenkins, J. C. (2012). Shifting public opinion on climate change: An empirical assessment of factors influencing concern over climate change in the U.S., 2002–2010. *Climatic Change, 114*(2), 169–188. https://doi.org/10.1007/s10584-012-0403-y

Brüggemann, M., & Engesser, S. (2017). Beyond false balance: How interpretive journalism shapes media coverage of climate change. *Global Environmental Change, 42*, 58–67. https://doi.org/10.1016/j.gloenvcha.2016.11.004

Carvalho, A. (2008). Communicating climate change in Portugal: A critical analysis of journalism and beyond. In A. Carvalho (Ed.), *Communicating climate change: Discourses, mediations and perceptions* (pp. 126–156). Braga: Centro de Estudos de Comunicação e Sociedade, Universidade do Minho.

Cissel, M. (2012). Media framing: A comparative content analysis on mainstream and alternative news coverage of Occupy Wall Street. *The Elon Journal of Undergraduate Research in Communications, 3*(1), 67–77.

Corner, A., Shaw, C., & Clarke, J. (2018). *Principles for effective communication and public engagement on climate change: A handbook for IPCC authors.* Oxford: Climate Outreach.

Davidson, K. (2020). Urgent need for post-"routine" climate strategies. *One Earth, 2*(2), 117–119. https://doi.org/10.1016/j.oneear.2020.01.014

Entman, R. M. (1993). Framing: Toward clarification of a fractured paradigm. *Journal of Communication, 43*(4), 51–58. https://doi.org/10.1111/j.1460-2466.1993.tb01304.x

European Commission. (2019). *The European Green Deal.* Communication from the Commission to the European Parliament, the European Council, the Council, the European Economic and Social Committee and the Committee of the Regions. Retrieved from https://ec.europa.eu.

González-Riancho, P., Gerkensmeier, B., & Ratter, B. M. (2017). Storm surge resilience and the Sendai Framework: Risk perception, intention to prepare and enhanced collaboration along the German North Sea coast. *Ocean & Coastal Management, 141*, 118–131. https://doi.org/10.1016/j.ocecoaman.2017.03.006

Gurevitch, M., & Blumler, J. (1990). Political communication systems and democratic values. In J. Lichtenberg (Ed.), *Democracy and the mass media: A collection of essays* (pp. 24–35). Cambridge: Cambridge University Press.

Hmielowski, J. D., Feldman, L., Myers, T. A., Leiserowitz, A., & Maibach, E. (2013). An attack on science? Media use, trust in scientists, and perceptions of global warming. *Public Understanding of Science, 23*(7), 866–883. https://doi.org/10.1177/0963662513480091

Horta, A., & Carvalho, A. (2017). Climate change communication in Portugal. In W. Leal Filho, E. Manolas, A. Azul, U. Azeiteiro, & H. McGhie (Eds.), *Handbook of climate change communication.* https://doi.org/10.1093/acrefore/9780190228620.013.599

Horta, A., Carvalho, A., & Schmidt, L. (2017). The hegemony of global politics: News coverage of climate change in a small country. *Society & Natural Resources, 30*(10), 1246–1260. https://doi.org/10.1080/08941920.2017.1295497

Jang, S. M. (2013). Framing responsibility in climate change discourse: Ethnocentric attribution bias, perceived causes, and policy attitudes. *Journal of Environmental Psychology, 36*, 27–36. https://doi.org/10.1016/j.jenvp.2013.07.003

Kahan, D. (2014). Making climate-science communication evidence-based—All the way down. In D. Crow, & M. Boykoff (Eds.), *Culture, politics and climate change: How information shapes our common future.* London: Routledge.

Karlsson-Vinkhuyzen, S. I., Friberg, L., & Saccenti, E. (2016). Read all about it!? Public accountability, fragmented global climate governance and the media. *Climate Policy, 17*(8), 982–997. https://doi.org/10.1080/14693062.2016.1213695

Moravcsik, A. (2004). Is there a 'democratic deficit' in world politics? A framework for analysis. *Government and Opposition, 39*(2), 336–363. https://doi.org/10.1111/j.1477-7053.2004.00126.x

Moser, S. C. (2010). Communicating climate change: History, challenges, process and future directions. *Wiley Interdisciplinary Reviews: Climate Change, 1*(1), 31–53. https://doi.org/10.1002/wcc.11

Moser, S. C., & Dilling, L. (2012). Communicating climate change: Closing the science–action gap. In J. Dryzek, R. Norgaard, & D. Schlosberg (Eds.), *The Oxford handbook of climate change and society* (pp. 161–173). New York, NY: Oxford University Press.

Nisbet, M. C. (2009). Communicating climate change: Why frames matter for public engagement. *Environment: Science and Policy for Sustainable Development, 51*(2), 12–23. https://doi.org/10.3200/envt.51.2.12-23

Nisbet, M. C., & Myers, T. (2007). The polls trends: Twenty years of public opinion about global warming. *Public Opinion Quarterly, 71*(3), 444–470. https://doi.org/10.1093/poq/nfm031

O'Brien, K. (2018). Is the 1.5°C target possible? Exploring the three spheres of transformation. *Current Opinion in Environmental Sustainability, 31,* 153–160. https://doi.org/10.1016/j.cosust.2018.04.010

Ockwell, D., Whitmarsh, L., & O'Neill, S. (2009). Reorienting climate change communication for effective mitigation. *Science Communication, 30*(3), 305–327. https://doi.org/10.1177/1075547008328969

Patt, A. G., & Weber, E. U. (2013). Perceptions and communication strategies for the many uncertainties relevant for climate policy. *Wiley Interdisciplinary Reviews: Climate Change, 5*(2), 219–232. https://doi.org/10.1002/wcc.259

Ripple, W. J., Wolf, C., Newsome, T. M., Barnard, P., & Moomaw, W. R. (2020). World scientists' warning of a climate emergency. *BioScience, 70*(1), 8–12. https://doi.org/10.1093/biosci/biz088

Romps, D. M., & Retzinger, J. P. (2019). Climate news articles lack basic climate science. *Environmental Research Communications, 1*(8), 081002. https://doi.org/10.1088/2515-7620/ab37dd

Smith, J. (2005). Dangerous news: Media decision making about climate change risk. *Risk Analysis, 25*(6), 1471–1482. https://doi.org/10.1111/j.1539-6924.2005.00693.x

Spence, A., & Pidgeon, N. (2009). Psychology, climate change and sustainable behaviour. *Environment, 51,* 8–18.

Spence, A., & Pidgeon, N. (2010). Framing and communicating climate change: The effects of distance and outcome frame manipulations. *Global Environmental Change, 20,* 656–667. https://doi.org/10.1016/j.gloenvcha.2010.07.002

Steffek, J. (2010). Public accountability and the public sphere of international governance. *Ethics & International Affairs, 24*(1), 45–68. https://doi.org/10.1111/j.1747-7093.2010.00243.x

Tavares, A. O., Areia, N. P., Mellett, S., James, J., Intrigliolo, D. S., Couldrick, L. B., & Berthoumieu, J. (2020). The European media portrayal of climate change: Implications for the social mobilization towards climate action. *Sustainability, 12*(20), 8300. https://doi.org/10.3390/su12208300

UN General Assembly. (2015). *Transforming our world: the 2030 Agenda for Sustainable Development.* A/RES/70/1. Retrieved from https://www.refworld.org/docid/57b6e3e44.html.

UNISDR (United Nations Office for Disaster Risk Reduction). (2015). *Sendai Framework for Disaster Risk Reduction 2015–2030.* Geneva: UNISDR.

Uzzell, D. L. (2019). The psycho-spatial dimension of global environmental problems. *Journal of Environmental Psychology, 20*(4), 307–318. https://doi.org/10.1006/jevp.2000.0175

Wolf, J., & Moser, S. C. (2011). Individual understandings, perceptions, and engagement with climate change: Insights from in-depth studies across the world. *Wiley Interdisciplinary Reviews: Climate Change, 2*(4), 547–569. https://doi.org/10.1002/wcc.120

Yu, T., Lin, F., Kao, K., Chao, C., & Yu, T. (2019). An innovative environmental citizen behavior model: Recycling intention as climate change mitigation strategies. *Journal of Environmental Management, 247,* 499–508. https://doi.org/10.1016/j.jenvman.2019.06.101

Investing in flood adaptation in Jakarta, Indonesia

Gusti Ayu Ketut Surtiari[1,2], Matthias Garschagen[3], José Manuel Mendes[4] and Yus Budiyono[5]

[1]*Research Center for Population, Indonesian Institute of Sciences, Jakarta, Indonesia;* [2]*University of Bonn, Bonn, Germany;* [3]*Department of Geography, Ludwig-Maximilians University of Munich, Munich, Germany;* [4]*Centre for Social Studies and Faculty of Economics, University of Coimbra, Coimbra, Portugal;* [5]*Agency for the Assessment and Application of Technology, Jakarta, Indonesia*

1. Evaluating adaptation to climate change

Climate change has been intensifying several hazards, regardless of their type and pace. Among these, floods stand out as particularly frequent extreme events in low-lying coastal areas and deltas, being exacerbated by drivers such as sea-level rise, land subsidence, and rapid urbanization (de Koning, Filatova, Need, & Bin, 2019; IPCC, 2018; Müller, 2013). Floods are also among the disasters that have been causing significant economic losses and damage across the globe and continue to challenge the achievement of the Sustainable Development Goals (IPCC, 2018). Although countries worldwide have responded to flood disasters with diverse policies and programs, floods' associated losses persist in high numbers (Djalante, Garschagen, Thomalla, & Shaw, 2017).

Building on the commitment ensuing from the Hyogo Framework for Action 2010–2015, the Sendai Framework for Disaster Risk Reduction 2015–2030 highlights, in its third priority, the importance of investing in disaster risk reduction (DRR) and resilience (UNISDR, 2015). As climate change increases the uncertainty related to both natural hazards and their impacts on social and ecological systems, it is essential to invest in DRR. Despite their differences in terms of addressed hazards,

Investing in Disaster Risk Reduction for Resilience. https://doi.org/10.1016/B978-0-12-818639-8.00006-5

timescales, and society—environment interactions (Kelman, Mercer, & Gaillard, 2017), DRR and climate change adaptation (CCA) are so interconnected that they should be pursued as a common goal. Kelman, Mercer, and Gaillard (2017) even claim that CCA is logically embedded in DRR. At the same time, some researchers contend that adaptation can be a timely option in responding to slow-onset events while ensuring long-term sustainability goals (Caravani, 2015; Ishiwatari & Surjan, 2019).

Being one of the megacities in Southeast Asia that are highly susceptible to the impacts of climate change, Jakarta has been implementing CCA measures in line with its development program. Due to the city's geographical features, floods in Jakarta were already recorded a century ago. However, the intensity and frequency of floods have increased significantly, leading to severe impacts on the population, businesses, and urban infrastructures. Floods currently represent an extreme event that is compounded by such factors as climate-change impacts, drainage-system deficiencies, and persistent land-subsidence problems. To account for these issues, the government enacted the regional adaptation planning in 2014. Yet, its implementation and evaluation still face several challenges. In particular, the review of the adaptation measures considering the standpoint of vulnerable communities is urgently needed because slum dwellers constitute the group that is most affected concomitantly by climate-change impacts and by formal adaptation measures. Indeed, the Indonesian capital has more than 600 *kampungs* (densely populated informal settlements), many of which located in flood-prone riverbanks, lakesides, and low-lying coastal areas (Simarmata & Surtiari, 2020). Nevertheless, the government initiatives being carried out to adapt coastal Jakarta to floods hardly include the point of view of these communities (Garschagen, Surtiari, & Harb, 2018).

This chapter attends to the third priority of the Sendai Framework by analyzing formal adaptation measures that ideally should increase the capacity for building resilience among vulnerable communities. Adaptation evaluation is a procedure to ensure the avoidance of maladaptation or unintended consequences of the existing measures. We propose an alternative evaluation framework to review formal adaptation measures by considering enabling factors within the adaptive capacity. Our analysis is based on two household surveys with slum dwellers living in coastal Jakarta whose areas were subject to formal adaptation measures.

This chapter is structured in five sections besides this one, in which we described the study's background. We introduce the alternative adaptation-evaluation framework in Section 2, before describing the case-study context in Section 3. In Section 4, we explain the research method, whereas in Section 5 we present the results and the ensuing discussion. This is followed, in Section 6, by the conclusions and an outlook for further studies.

2. Considering enabling factors to evaluate adaptation

According to the IPCC (2018), adaptation is an adjustment in a natural and human system to respond to actual and projected climate change while also providing benefits for it to be more resilient. Adaptation can thus be an opportunity to tackle vulnerability drivers and also enhance the capacity of a system to build resilience. However, adaptation can create other risks in the future, especially if it fails to respond to climate change's long-term impacts, a situation that is generally defined as maladaptation (Hallegatte, Vogt-Schilb, Bangalore, & Rozenberg, 2017; Juhola, Glaas, Linnér, & Neset, 2016). To ensure that adaptation initiatives will effectively respond to the identified issues, it is essential to duly assess CCA projects before their implementation and then to evaluate their effectiveness retrospectively. However, adaptation evaluation is a challenging endeavor given the involved long-term horizon to ensure that the initiatives are truly successful.

This is especially the case of formal adaptation—that is, interventions deliberately carried out by governmental actors (Birkmann et al., 2010)—whose outcomes and drawbacks cannot be fully understood a priori. For instance, the question of how vulnerable groups could benefit from formal adaptation is still under consideration, namely regarding its complex impacts on vulnerability causes. In this regard, Garschagen (2014) showed that adaptation can indirectly influence vulnerability depending on the adaptive capacity's building and consolidation. Following Grothmann and Patt (2005), we consider adaptive capacity here as the accumulation of productive capital coupled with social and cognitive aspects that can affect the success of adaptation in the future.

Our proposal of evaluating adaptation is based on an integrative framework that combines adaptation and vulnerability, the latter being defined as the degree of a system to be harmed by hazards (IPCC, 2012). This framework considers enabling factors that can influence the vulnerability of affected communities. To develop it, we considered two approaches. The first one was the actor-oriented and context-specific framework proposed by Krause, Schwab, and Birkmann (2015). This approach highlights the actors' subjective judgments to reduce vulnerability based on their self-evaluation (Grothmann & Patt, 2005). Risk perception and its changes over time represent a key variable to evaluate vulnerability. The second approach was Garschagen's advanced integrated framework for vulnerability and adaptation analysis (Garschagen, 2014). Our integrative adaptation-evaluation framework considers the dynamics of vulnerability induced by adaptation measures that aimed to lower existing vulnerability levels. To this end, we consider that the impact of adaptation measures on vulnerability indicators depends on sociocognitive factors related to the targeted vulnerable groups, which consist in their capacity to anticipate and cope with future climate-change risks.

In CCA research, the vulnerability concept has a set of core factors, namely sensitivity and coping (or adaptive) capacity of an individual, a group, or a system to reduce the impacts of climate change (Birkmann, 2013). Thus, vulnerability considers the adaptive capacity within the system, albeit it is also composed of two other components: exposure—the intensity of the occupation of a hazard-prone zone—and sensitivity—the degree of fragility of a given population (Birkmann, 2013). The latter is fed by structural conditions such as inequality, marginalization, and poverty that disproportionately impinge on certain social groups. The linkages between adaptation and vulnerability indicators occur by causing opportunities or barriers to enhance people's capacity to adapt (Garschagen, 2014). It means that one can judge the impacts of adaptation measures across scales and levels during their implementation process (Adger, Arnell, & Tompkins, 2005; Moser & Boykoff, 2013). For instance, the government's adaptation measures and policies can cause direct and/or indirect impacts on vulnerable groups, particularly in areas prone to climate-related hazards.

Vulnerability to climate change is dynamic over time and scales (Birkmann, 2013). To assess the dynamics of vulnerability, existing studies apply at least three approaches. The first group adopts a historical approach, focusing on two aspects: the analysis of social-vulnerability patterns based on the sociodemographic indicators extracted from population census data (Kashem, Wilson, & Van Zandt, 2016), or the analysis of exposure changes over time (Kuhlicke, 2010). The second follows a time-dimension approach, for instance, by observing the distribution of people at risk in the morning, afternoon, and at night (Setiadi, 2014), or by analyzing the vulnerability before and after adaptation interventions (Birkmann, 2013). The third follows a group-category approach, which analyzes the impact of the same shocks on different people (Leichenko & O'Brien, 2002).

Following the second group, changes in the degree of vulnerability of a given community constitute a timely indicator to evaluate adaptation, but these can only be confirmed after several shocks or the occurrence of the projected extreme events. However, vulnerability can also be analyzed as an outcome by considering the capacity to adapt as a means leading to vulnerability reduction. On the other hand, vulnerability reduction depends on the willingness to take action and also on the capacity to manage resources to respond to risks (Garschagen, 2014; Pelling, 2011). Adaptations that aim to respond to climate change can influence other capacities of vulnerable groups, even though they have no intention to explicitly address them. The first order of adaptation can create a new environment to which the surrounding inhabitants need to adjust. The limited capacity to adjust in the second stage of the changes can lead a system into new risks; anticipating cascading processes is thus an essential aspect in understanding adaptation to climate change (Birkmann, 2011).

To analyze the dynamics of vulnerability as an approach to evaluating adaptation, one needs to identify the links between vulnerability and formal adaptation. To this end, we applied the concept of adaptation as an action (Eisenack & Stecker, 2012), in which we consider that government acts as an operator, and affected communities are the recipient of the impacts of such formal adaptations. Adaptation failure can happen when the operators cannot provide robust and appropriate strategies to reduce risks, or the recipients do not succeed to adjust to the changes caused by the measures proposed by the operators. However, based on Giddens's structuration theory, individuals are not passive and, as actors, have their own rationale and strategies to respond to risks (Fuchs, 2003; Thompson, 2012). At the same time, the decision to take action and the willingness to enhance or transform one's capacity to adapt rely on how people perceive risks, which is constructed by their knowledge and experiences (Pelling, 2011).

Moreover, self-efficacy leads to specific actions to respond to expected climate-change impacts (Grothmann & Patt, 2005). Therefore, considering perception issues within adaptation evaluation is essential regarding the potential impacts and opportunities from formal adaptation processes that can benefit the affected communities. Also, the willingness to take adaptive action is influenced by institutional processes. In the case of slums and urban poor, the informal system plays a significant role to build resilience (Simarmata, 2018). By living under unfavorable conditions on a daily basis, the urban poor have established their informal adaptation as means to respond to risks and to put their efforts to adjust to environmental changes (Simarmata, 2018). In the occurrence of a disaster, informal systems are crucial to provide early warning and first response efforts, as well as to sustain recovery (van Voorst, 2016). When looking for assistance during and after a disaster, urban dwellers often prefer to resort to informal institutions based on trust and acceptable procedures, rather than going through formal institutions and their particular bureaucracy (van Voorst, 2016).

In the proposed framework (Fig. 11.1), the vulnerability changes caused by the implementation of formal adaptation measures will depend on the enabling factors of the adaptive

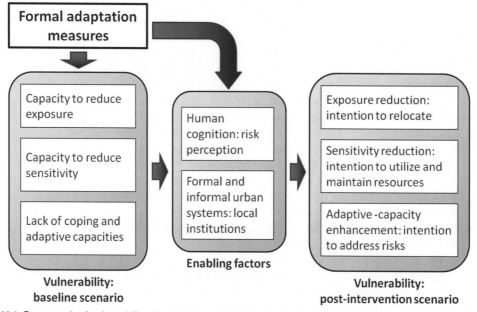

Figure 11.1 Framework of vulnerability dynamics for adaptation evaluation. *Based on* Birkmann, J., Cardona, O. D., Carreño, M. L., Barbat, A. H., Pelling, M., Schneiderbauer, S., … Welle, T. (2013). Framing vulnerability, risk and societal responses: The MOVE framework. *Natural Hazards, 67*(2), 193—211. https://doi.org/10.1007/s11069-013-0558-5, Eisenack, K., & Stecker, R. (2012). A framework for analyzing climate change adaptations as actions. *Mitigation and Adaptation Strategies for Global Change, 17*(3) 243—260. https://doi.org/10.1007/s11027-011-9323-9; Garschagen, M. (2014). *Risky change? Vulnerability and adaptation between climate change and transformation dynamics in Can Tho City*, Vietnam. Stuttgart: Franz Steiner Verlag; Grothmann, T., & Patt, A. (2005). Adaptive capacity and human cognition: The process of individual adaptation to climate change. *Global Environmental Change, 15*(3), 199—213. https://doi.org/10.1016/j.gloenvcha.2005.01.002, and IPCC (Intergovernmental Panel on Climate Change). (2012). *Managing the risks of extreme events and disasters to advance climate change adaptation. A special report of Working Groups I and II.* Cambridge: Cambridge University Press.

response, which are shaped by risk perception and the institutions within the system. Risk perception in this study is not merely the self-interpretation of an individual judgment but, instead, the integrated understanding of risks' causes and consequences (Buchecker et al., 2013). Therefore, risk perception not only is limited to flooding but also includes grasping other risks engendered by the new social and physical environments ensuing from formal adaptation measures.

3. Context of the study: Jakarta, Indonesia

Located on the north coast of Java Island, Jakarta is inhabited by almost 12 million people, corresponding to about 4% of Indonesia's total population (BPS Jakarta, 2020). In its turn, Greater Jakarta includes the surrounding cities of Depok, Bogor, Tangerang, and Bekasi and has a population of about 29.3 million (BPS Jakarta, 2019). In the river-basin context, Jakarta is located in a delta, whereas the surrounding cities—particularly Bogor—are mainly upstream of the rivers flowing into the Indonesian capital. This geographical context is quite complex, as Jakarta is crossed by 13 main rivers and 2 canals (Fig. 11.2). Located along at least

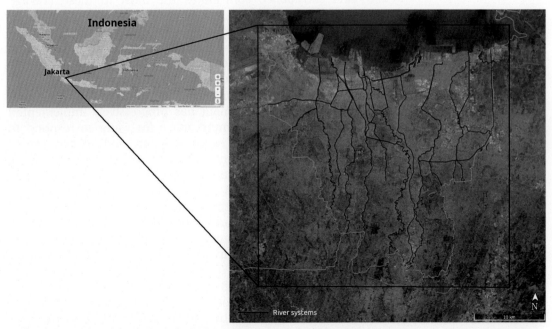

Figure 11.2 Jakarta's main hydrological networks. Adapted from GIS BPBD Jakarta on top of a *Google Earth* image.

one of these waterways, 73% of Jakarta's subdistricts are prone to floods (BPS Jakarta, 2020).

North Jakarta is one of Indonesia's most vulnerable districts regarding climate-change impacts, particularly sea-level rise and increasing frequency and intensity of precipitations (Firman, Surbakti, Idroes, & Simarmata, 2011; Yusuf & Francisco, 2009). Climate change, interlinked with urbanization, land subsidence, and insufficient drainage system, has been triggering more severe floods in the past decades (Abidin et al., 2011; Budiyono, Aerts, Brinkman, Marfai, & Ward, 2015; Chaussard, Amelung, Abidin, & Hong, 2013). The number of informal housings is uncontrolled and grows fast given that the informal sectors play an important role in particular areas of the city (World Bank, 2011a). For instance, the northern part of Jakarta has a busy fishing port, which boosts the related social and economic activities in a fragile environmental setting (Surtiari, Djalante, Setiadi, & Garschagen, 2017). These factors in combination have been increasing the risk of extreme floods, both tidal and urban ones.

Land subsidence also aggravates this scenario and, coupled with climate change, has been causing floods whose impacts are hard to predict (Budiyono, 2018; Garschagen et al., 2018). The projection of coastal floods in 2050 shows that Jakarta will face severe problems, including those related to sea-level rise, and land subsidence contributes the most to these (Takagi, Esteban, Mikami, & Fujii, 2016). Climate change has been increasing rainfall intensity, as exemplified by the flash floods that occurred at the beginning of 2020 (Rahmayanti, Azzahra, & Arnanda, 2021). The impacts can be even worse and unpredictable if there are no significant adaptive and appropriate strategies in place to manage the drainage system (Budiyono, 2018).

Since 2012, the government has progressively implemented the concept of providing more space for water to respond to the increasing impact of extreme floods and related causes (particularly land subsidence) in Jakarta. This implies carrying out intensive formal adaptation interventions such as building coastal dykes, widening and embanking rivers, rehabilitating water reservoirs, and relocating inhabitants—slum dwellers—of flood-prone riverbanks (World Bank, 2011b). Given that space is a scarce resource in this dense metropolis, the implementation of these measures always involves important trade-offs (Simarmata & Surtiari, 2020).

Supported by the World Bank, the hard-engineering urban adaptations are located in poor urban neighborhoods of coastal Jakarta, mostly inhabited by migrants who work in informal and low-skilled jobs in the fishery sector (Garschagen et al.,

2018; Surtiari et al., 2017). By relocating people to safer areas, the program intended to provide more space for both protective infrastructures and water during floods, as well as to reduce the number of slums, while providing its inhabitants with better living conditions. Yet, the program's implementation was fraught with many challenges, particularly for relocating the vulnerable communities. Issues related to social justice, sustained livelihoods, or disrupted social ties, for instance, did not encourage the acceptance of prompt relocation. Yet, the devastating 2013 flood forced hundreds of families to be relocated and also led to the construction of a coastal dyke next to an urban-poor settlement (Simarmata & Surtiari, 2020).

4. The study's methodology

This study explored vulnerability dynamics among slum dwellers as a means to evaluate CCA initiatives. To this end, we analyzed the changes in enabling factors that were induced by the government's implementation of formal adaptation measures and policies. First, our analysis focused on the changes in people's perception of the official programs and of their self-capacity to manage future risks. Secondly, we analyzed the changes that occurred in local communities' institutions, both formal and informal ones, as significant drivers to manage future risks. We consider these—risk perception and local institutions—to be enabling factors to reduce both flood vulnerability and exposure and to efficiently utilize resources to increase adaptive and coping capacities.

The study followed a mixed-method—qualitative and quantitative—approach and sought in-depth and comprehensive data on vulnerability changes in local communities. The combined approach aimed to collect complementary data, which was eventually subject to mixed analysis (Yin, 2014). To gather data, we conducted household surveys in 2015 and 2017, as well as semistructured in-depth interviews with selected respondents: local leaders and inhabitants who have been living for at least four decades in the study areas. Through such data triangulation, we sought to deepen the information related to the socio-cognitive aspects.

The study areas are located along the coast and were selected purposively based on the particular infrastructures developed to manage extreme flood risk. The four studied slum sites are exposed to tidal and urban floods and directly relate to the recently built formal adaptation infrastructures. The 451 survey respondents were distributed almost equally across the four neighborhoods. We selected the respondents based on their living location and the fact that their informal housing was affected by a formal adaptation measure.

The 2015 household survey identified the impact of the coastal dyke, the reservoir, and the relocation of affected persons one to two years after their implementation. In 2017, the same respondents were again contacted to evaluate their conditions three to four years after living under the new conditions brought by these adaptation measures. We also conducted semistructured interviews and focus-group discussions between the two household surveys to complement the analysis at the individual level with a more collective standpoint. We analyzed the data ensuing from the two surveys through descriptive statistics, whereas for the data gathered through interviews and discussions we coded, categorized, sorted, and then interpreted the information.

5. Enabling adaptation in coastal Jakarta

In the next subsections, we analyze two leading indicators within the evaluation process: first, the changes in risk perception caused by the implementation of adaptation measures; and second, the changes incurred in the local institutions of the vulnerable communities.

5.1 Changes in risk perception: impressions about flood safety

Following Pelling (2011), we took as a baseline that risk perception influences people's willingness to take action. Thus, we examined how the formal adaptation measures implemented by the government impacted the way that the affected communities perceived risks.

In general, the vulnerable slum dwellers did not perceive extreme floods as a risk but rather as a regular event. However, the increased intensity and frequency of floods, as experienced in 2007 and 2013, have raised their concern and awareness about the risks related to extreme events. The perception of the adaptation measures is significantly different among the four communities. Overall, slum dwellers stated that these measures aim to reduce flood-risk drivers, but in a broader understanding, they mentioned also the aim of protecting the coastal area. They also understand that it is not sufficient enough to tackle all flood-prone coastal areas and particularly that these adaptations do not favor the urban poor. Indeed, for those living in a flood-prone area, the root causes of flooding still persist and challenge them to accomplish successful informal adaptations. Thus, they keep considering putting forward their spontaneous strategies to anticipate and cope with disaster risk.

In the case of the communities targeted by the reservoir and relocation measures, the respondents stated that these adaptations are useful mainly to cope with the risk of flooding. However, almost half of the respondents affected by the dyke construction indicated that the program is not affirmative. This perception correlates with several other variables, such as the knowledge about climate change, negative impact on the households, level of education, changing housing status, social cohesion, flood exposure, risk perception on floods, and income (correlation is significant at the 1% level). The perception of the adaptation measure also correlates with people's livelihoods and the flood experience after its implementation (correlation is significant at the 5% level). Fig. 11.3 shows the verified changes in risk

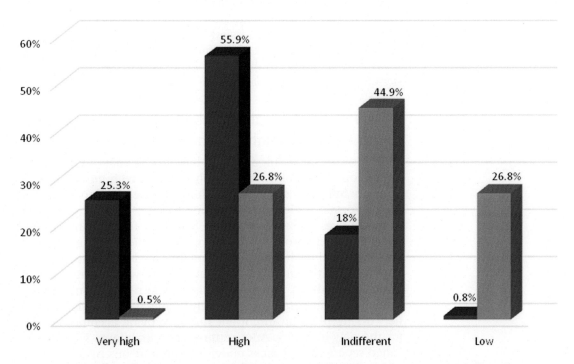

Figure 11.3 Changes in risk perception to coastal floods between 2015 and 2017, as revealed by the household surveys (N = 451). The authors.

perception—slum dwellers agree that the implemented adaptation measures are helpful and coastal floods will generally decrease. This changed perception was particularly notable among the relocated people.

Moreover, the formal adaptation measures also indirectly influenced risk perception among the affected communities. This trend is evidenced by the respondents' evaluation of the effectiveness of the formal adaptations. Overall, around 60% of the respondents had a positive evaluation of the program consisting of reservoirs, dykes, and relocations. The respondents stated that flood risk might decrease in the future as several measures have been put in place at once. Yet, they also perceived that these measures might not provide significant benefits for them as slum dwellers. They agreed that the coastal dyke, the reservoir normalization, and other measures contribute to reducing floods in the city. Nevertheless, they perceive that the development process associated with these adaptation measures is marked by inequalities. The interviews showed that whereas the program aims to reduce the number of people exposed in the flood-prone areas by relocating slum dwellers, at the same time several exclusive residential and business areas are under construction in these areas. Accordingly, the interviewees affected by the construction of the coastal dyke agree that this measure will not manage to solve the flooding problem as long as the construction of skyscrapers along the coast is still prominent.

5.2 Changes in local institutions: from an informal to a formal neighborhood system

The second enabling factor to link formal adaptation and its impact on vulnerability is the evolvement of community institutions in informal settlements. Slum dwellers understand that their informal settlement has a long history, which started with the initiatives of several families to utilize flood-prone coastal areas to benefit from the booming economic activities nearby. This implies that the number of inhabitants in the studied areas has been continuously growing, in tandem with the number of informal buildings. Since their origins, the negotiation of living spaces has been arranged informally among the pioneers; kinship networking plays an important role to provide accommodation for the new migrants. After settling, the inhabitants have established secure living spaces under the spontaneous neighborhood arrangement. We also identified such self-organization processes in the government's distribution of social assistance, targeting several registered poor families. The community leader often redistributes such assistance to the families most in need. More-over, after a disaster, all the people help each other during the re-covery stage.

Our study found that the formalization of the existing neigh-borhood system does not automatically upgrade the lives of the affected communities. The household surveys showed that not all respondents benefit from the interventions provided by the local government. In the case of relocation, the government pro-vided free renting for six months and free public transportation from the new vertical housing to work, as well as training oppor-tunities. But these incentives did not straightforwardly support a transition from an informal to a formal system.

For instance, the relocated families faced greater challenges regarding income generation. Instead of increased confidence to face future floods, they got more concerned about the likely weakening of their sociocultural networks and relationships, which constitute their main assets to spontaneously manage risks in their neighborhood. Indeed, the relocation has changed the preexisting neighborhood system, namely the registration and rent-payment procedures followed by the inhabitants. The cur-rent monthly payment for renting a house is not supported by the common livelihood options in informal sectors, which are more flexible, thus uncertain regarding income sustainability. In the relocated families' previous informal settlement, rent payments were also flexible, based on mutual trust and under-standing between property landowners and them. The strict rules of the adopted administrative procedures along the relocation process have placed the affected communities at a new risk related to eviction, given the uncertainty of maintaining a place for living.

Moreover, the respondents also stated that social cohesion among community members has declined due to the emergence of new power relations between the relocated families and local (formal and informal) leaders. Yet, power relations among slum dwellers condition their access to information about the inter-ventions, including regarding compensation in case of eviction from the informal settlement. The interviews showed that the respondents have different levels of information on the process and the program's whole plan. Although the leading agency and institutions that are responsible for the relocation have disseminated the program, the views of the relocated inhabitants were only captured through the community representative. The process was not properly inclusive. As a consequence, the infor-mation among community members has been unbalanced while depending on formal and informal sources. For instance, the information among the affected communities often comes from online news or informal talks among the neighborhoods. The coexistence of competing sources of information negatively affects trust among community members and also between slum dwellers and local leaders. One of the major concerns is about the compensation from the local government to the affected communities. Some of the families stated that they received some compensation, whereas others highlighted that there is no compensation regarding the status of their land and housing.

6. Adapting formal climate change adaptation to meet the specificities of informality

In this chapter, we proposed an alternative approach to recognize the impacts of adaptation measures on building resilience. This approach also intended to understand how formal adaptation effectively reduces vulnerability while increasing the capacity to build resilience, particularly among vulnerable communities. To this end, we focused on enabling factors to link adaptation and the capacity to reduce vulnerability, namely risk perception and institutional changes. Our study shows that sudden changes and unintended consequences of the adaptation measures on the affected communities can hamper the maintenance and enhancement of their capacity to respond to new risks in the future.

In line with the adaptation concept, a successful intervention should provide opportunities for the targeted communities to benefit from the adopted strategies at any level. When it causes barriers to adjusting, the strategy should be evaluated and reoriented during its implementation. For the identification of successful adaptation, when it comes to slow-onset hazards—like the ones related to climate change—one should consider a long time horizon, and it would be helpful if the goals could be divided into short- and long-term scales (Moser & Ekstrom, 2010). Long-term adaptation requires a paradigm shift to transform a system to robustly anticipate climate change. Moreover, in the medium term, the adaptation should incite changes in key aspects that can lead to a system's complete transformation (Moser & Ekstrom, 2010).

Instead of evaluating the decision-making process related to an adaptation plan, the approach developed in this chapter analyzes current drivers shaping vulnerability. It provides a preliminary description for long-term vulnerability considering human cognition, particularly the effects of formal adaptation measures on risk perception. Besides, adaptation's institutional processes and their relation to risk perception can influence the decisions to take adaptive action, such as the intention to move away from risk-prone areas. The self-capacity perception is also an essential factor in enhancing adaptive capacity (Grothmann & Patt, 2005). This study shows that formal adaptation more likely influences the change of risk perception and, therefore, affects new risks in the future.

There is a pitfall in formal adaptation when it aims to institutionalize structures and neglects informality. Our study significantly illustrated that informal systems are important for reducing vulnerability, expressed, for instance, by their flexibility to cope with and anticipate economic pressures caused by formal adaptation. The continuity of subsistence economic activities is a big challenge in the future if the appropriate assistance and intervention measures are lacking. As adaptation is a dynamic process, it continuously influences the social structure of the affected communities. This study showed that when the government tries to formalize local institutions, the community's social structure tends to lose part of its ability to enhance adaptive capacity. Therefore, informality should be considered as having positive contributions for effective formal adaptation. Consequently, the success of formal adaptation depends on urban dwellers' involvement in adaptation planning and implementation.

By presenting a CCA evaluation framework based on the dynamics of vulnerability, this chapter put in evidence some of the contradictions involved in CCA and DRR, which are overlooked in the Sendai Framework. As sagaciously perceived by the slum dwellers consulted during this study, infrastructural solutions focused on hazard mitigation may not be enough to handle future flood risks. Accordingly, one may question to what extent hazard mitigation can be truly qualified as an adaptation measure, as it fails to provide co-benefits in terms of building social resilience. This lesson calls for the need to invest also to address the vulnerabilities of the exposed people, by improving their adaptive capacity and reducing their fragility—namely through more inclusive policies, plans, and strategies to tackle inequality and poverty. This implies both the full integration of social-justice considerations into adaptation and the recognition of informality's contributions to resilience building. Taking these two conditions into account is crucial when evaluating adaptation.

Acknowledgments

This study is part of the first author's PhD research, funded by the German Ministry of Education and Research under the TWIN-SEA project at the United Nations University Institute for Environment and Human Security (UNU-EHS). This chapter further develops the paper that the first author presented in 2018 at the 8th International Conference on Building Resilience, in Lisbon, with the support of a grant from the Indonesian Institute for Sciences.

References

Abidin, H. Z., Andreas, H., Gumilar, I., Fukuda, Y., Pohan, Y. E., & Deguchi, T. (2011). Land subsidence of Jakarta (Indonesia) and its relation with urban development. *Natural Hazards, 59*(3), 1753–1771. https://doi.org/10.1007/s11069-011-9866-9

Adger, W. N., Arnell, N. W., & Tompkins, E. L. (2005). Successful adaptation to climate change across scales. *Global Environmental Change, 15*(2), 77–86. https://doi.org/10.1016/j.gloenvcha.2004.12.005

Birkmann, J. (2011). First- and second-order adaptation to natural hazards and extreme events in the context of climate change. *Natural Hazards, 58*(2), 811–840. https://doi.org/10.1007/s11069-011-9806-8

Birkmann, J. (2013). Measuring vulnerability to promote disaster-resilient societies and to enhance adaptation: Conceptual frameworks and definitions. In J. Birkmann (Ed.), *Measuring vulnerability to natural hazards: Towards disaster resilient societies* (2nd ed., pp. 9–79). Tokyo: UNU Press.

Birkmann, J., Buckle, P., Jaeger, J., Pelling, M., Setiadi, N., Garschagen, M., ... Kropp, J. (2010). Extreme events and disasters: A window of opportunity for change? Analysis of organizational, institutional and political changes, formal and informal responses after mega-disasters. *Natural Hazards, 55*(3), 637–655. https://doi.org/10.1007/s11069-008-9319-2

Birkmann, J., Cardona, O. D., Carreño, M. L., Barbat, A. H., Pelling, M., Schneiderbauer, S., ... Welle, T. (2013). Framing vulnerability, risk and societal responses: The MOVE framework. *Natural Hazards, 67*(2), 193–211. https://doi.org/10.1007/s11069-013-0558-5

BPS Jakarta. (2019). *The province of Jakarta in figures 2019*. Jakarta, Indonesia.

BPS Jakarta. (2020). *The province of Jakarta in figures 2020*. Jakarta, Indonesia.

Buchecker, M., Salvini, G., Di Baldassarre, G., Semenzin, E., Maidl, E., & Marcomini, A. (2013). The role of risk perception in making flood risk management more effective. *Natural Hazards and Earth System Sciences, 13*(11), 3013–3030. https://doi.org/10.5194/nhess-13-3013-2013

Budiyono, Y. (2018). *Flood risk modeling in Jakarta: Development and usefulness in a time of climate change*. Amsterdam: Vrije Universiteit Amsterdam.

Budiyono, Y., Aerts, J., Brinkman, J., Marfai, M. A., & Ward, P. (2015). Flood risk assessment for delta mega-cities: A case study of Jakarta. *Natural Hazards, 75*(1), 389–413. https://doi.org/10.1007/s11069-014-1327-9

Caravani, A. (2015). *Does adaptation finance invest into disaster risk reduction? ODI report*. Retrieved from https://cdn.odi.org/media/documents/9766.pdf.

Chaussard, E., Amelung, F., Abidin, H., & Hong, S.-H. (2013). Sinking cities in Indonesia: ALOS PALSAR detects rapid subsidence due to groundwater and gas extraction. *Remote Sensing of Environment, 128*, 150–161. https://doi.org/10.1016/j.rse.2012.10.015

de Koning, K., Filatova, T., Need, A., & Bin, O. (2019). Avoiding or mitigating flooding: Bottom-up drivers of urban resilience to climate change in the USA. *Global Environmental Change, 59*, 101981. https://doi.org/10.1016/j.gloenvcha.2019.101981

Djalante, R., Garschagen, M., Thomalla, F., & Shaw, R. (2017). Introduction. In R. Djalante, M. Garschagen, F. Thomalla, & R. Shaw (Eds.), *Disaster risk reduction in Indonesia: Progress, challenges, and issues* (pp. 1–17). Cham: Springer. https://doi.org/10.1007/978-3-319-54466-3_1

Eisenack, K., & Stecker, R. (2012). A framework for analyzing climate change adaptations as actions. *Mitigation and Adaptation Strategies for Global Change, 17*(3), 243–260. https://doi.org/10.1007/s11027-011-9323-9

Firman, T., Surbakti, I. M., Idroes, I. C., & Simarmata, H. A. (2011). Potential climate-change related vulnerabilities in Jakarta: Challenges and current status. *Habitat International, 35*(2), 372–378. https://doi.org/10.1016/j.habitatint.2010.11.01

Fuchs, C. (2003). Structuration theory and self-organization. *Systemic Practice and Action Research, 16*(2), 133–167. https://doi.org/10.1023/A:1022889627100

Garschagen, M. (2014). *Risky change? Vulnerability and adaptation between climate change and transformation dynamics in Can Tho City, Vietnam.* Stuttgart: Franz Steiner Verlag.

Garschagen, M., Surtiari, G. A. K., & Harb, M. (2018). Is Jakarta's new flood risk reduction strategy transformational? *Sustainability, 10*, 2934. https://doi.org/10.3390/su10082934

Grothmann, T., & Patt, A. (2005). Adaptive capacity and human cognition: The process of individual adaptation to climate change. *Global Environmental Change, 15*(3), 199–213. https://doi.org/10.1016/j.gloenvcha.2005.01.002

Hallegatte, S., Vogt-Schilb, A., Bangalore, M., & Rozenberg, J. (2017). *Unbreakable: Building the resilience of the poor in the face of natural disasters.* Washington, DC: World Bank. Retrieved from https://openknowledge.worldbank.org/handle/10986/25335.

IPCC (Intergovernmental Panel on Climate Change). (2012). *Managing the risks of extreme events and disasters to advance climate change adaptation. A special report of Working Groups I and II.* Cambridge: Cambridge University Press.

IPCC (Intergovernmental Panel on Climate Change). (2018). *Special report: Global warming of 1.5°C—Summary for policymakers.* Retrieved from https://www.ipcc.ch/site/assets/uploads/sites/2/2019/05/SR15_SPM_version_report_LR.pdf.

Ishiwatari, M., & Surjan, A. (2019). Good enough today is not enough tomorrow: Challenges of increasing investments in disaster risk reduction and climate change adaptation. *Progress in Disaster Sciences, 1*, 100007. https://doi.org/10.1016/j.pdisas.2019.100007

Juhola, S., Glaas, E., Linnér, B.-O., & Neset, T.-S. (2016). Redefining maladaptation. *Environmental Science and Policy, 55*(1), 125–140. https://doi.org/10.1016/j.envsci.2015.09.014

Kashem, S. B., Wilson, B., & Van Zandt, S. (2016). Planning for climate adaptation: Evaluating the changing patterns of social vulnerability and adaptation challenges in three coastal cities. *Journal of Planning Education and Research, 36*(3), 304–318. https://doi.org/10.1177/0739456X16645167

Kelman, I., Mercer, J., & Gaillard, J. C. (2017). Editorial introduction to this handbook: Why act on disaster risk reduction including climate change adaptation. In I. Kelman, J. Mercer, & J. C. Gaillard (Eds.), *The Routledge handbook of disaster risk reduction including climate change adaptation* (pp. 3–8). Abington: Routledge.

Krause, D., Schwab, M., & Birkmann, J. (2015). An actor-oriented and context-specific framework for evaluating climate change adaptation. *New Directions for Evaluation, 147*, 37–48. https://doi.org/10.1002/ev.20129

Kuhlicke, C. (2010). The dynamics of vulnerability: Some preliminary thoughts about the occurrence of "radical surprises" and a case study on the 2002 flood (Germany). *Natural Hazards, 55*(3), 671–688. https://doi.org/10.1007/s11069-010-9645-z

Leichenko, R. R. M., & O'Brien, K. L. K. (2002). The dynamics of rural vulnerability to global change: The case of southern Africa. *Mitigation and Adaptation Strategies for Global Change, 7*(1), 1–18. https://doi.org/10.1023/A:1015860421954

Moser, S. C., & Boykoff, M. T. (2013). *Successful adaptation to climate change: Linking science and policy in a rapidly changing world.* London: Routledge.

Moser, S. C., & Ekstrom, J. A. (2010). A framework to diagnose barriers to climate change adaptation. *PNAS, 107*(51), 22026–22031. https://doi.org/10.1073/pnas.1007887107

Müller, A. (2013). Flood risks in a dynamic urban agglomeration: A conceptual and methodological assessment framework. *Natural Hazards, 65*(3), 1931–1950. https://doi.org/10.1007/s11069-012-0453-5

Pelling, M. (2011). *Adaptation to climate change: From resilience to transformation.* Abingdon: Routledge.

Rahmayanti, K. P., Azzahra, S., & Arnanda, N. A. (2021). Actor-network and non-government failure in Jakarta flood disaster in January 2020. *IOP Conference Series: Earth and Environmental Science, 716*, 012053. https://doi.org/10.1088/1755-1315/716/1/012053

Setiadi, N. J. (2014). *Assessing people's early warning response capability to inform urban planning interventions to reduce vulnerability to tsunamis. Case Study of Padang City, Indonesia.* Doctoral thesis. Bonn: Rheinischen Friedrich-Wilhelms-Universität. Retrieved from https://bonndoc.ulb.uni-bonn.de/xmlui/bitstream/handle/20.500.11811/5829/3502.pdf.

Simarmata, H. A. (2018). Locally embedded adaptation planning. In H. A. Simarmata (Ed.), *Phenomenology in adaptation planning: An empirical study of flood-affected people in Kampung Muara Baru Jakarta* (pp. 105–143). Singapore: Springer.

Simarmata, H. A., & Surtiari, G. A. K. (2020). *Adaptation to climate change decision making and opportunities for transformation in Jakarta, Indonesia.* Research Paper 2020-3. Retrieved from https://www.unrisd.org/80256B3C005BCCF9/(httpAuxPages)/FA1FD6C0607E095380258639005271FA/$file/JAKARTA—Transformative-Adaptation-Coastal-Cities-2020.pdf.

Surtiari, G. A. K., Djalante, R., Setiadi, N. J., & Garschagen, M. (2017). Culture and community resilience to flooding: Case study of the urban coastal community in Jakarta. In R. Djalante, M. Garschagen, F. Thomalla, & R. Shaw (Eds.), *Disaster risk reduction in Indonesia: Progress, challenges, and issues* (pp. 469–493). Cham: Springer.

Takagi, H., Esteban, M., Mikami, T., & Fujii, D. (2016). Projection of coastal floods in 2050 Jakarta. *Urban Climate, 17*, 135–145. https://doi.org/10.1016/j.uclim.2016.05.003

Thompson, M. (2012). People, practice, and technology: Restoring Giddens' broader philosophy to the study of information systems. *Information and Organization, 22*(3), 188–207. https://doi.org/10.1016/j.infoandorg.2012.04.001

UNISDR. (2015). *Sendai Framework for Disaster Risk Reduction 2015–2030.* Retrieved from https://www.preventionweb.net/files/43291_sendaiframework fordrren.pdf.

van Voorst, R. (2016). Formal and informal flood governance in Jakarta, Indonesia. *Habitat International, 52,* 5–10. https://doi.org/10.1016/j.habitatint.2015. 08.023

World Bank. (2011a). *Jakarta—Urban challenges in a changing climate.* Retrieved from http://documents.worldbank.org/curated/en/132781468039 870805/Jakarta-Urban-challenges-in-a-changing-climate.

World Bank. (2011b). *Indonesia—Jakarta Urgent Flood Mitigation Project.* Project appraisal document. Retrieved from http://documents.worldbank.org/ curated/en/622381468263088609/Indonesia-Jakarta-Urgent-Flood-Mitigation-Project.

Yin, R. K. (2014). *Case study research: Design and methods* (5th ed.). Los Angeles, CA: Sage.

Yusuf, A. A., & Francisco, H. (2009). *Climate change vulnerability mapping for Southeast Asia.* Retrieved from https://www.preventionweb.net/publications/ view/7865.

Conclusion

Moving from frameworks to action: The importance of context-driven investments to deal with disasters' root causes

A. Nuno Martins[1], Gonzalo Lizarralde[2], Temitope Egbelakin[3], Liliane Hobeica[4], José Manuel Mendes[5] and Adib Hobeica[6]

[1] CIAUD, Research Centre for Architecture, Urbanism and Design, Faculty of Architecture, University of Lisbon, Lisbon, Portugal; [2] École d'Architecture, Université de Montréal, Montreal, Quebec, Canada; [3] School of Architecture and Built Environment, University of Newcastle, Newcastle, NSW, Australia; [4] RISKam (Research Group on Environmental Hazard and Risk Assessment and Management), Centre for Geographical Studies, University of Lisbon, Lisbon, Portugal; [5] Centre for Social Studies and Faculty of Economics, University of Coimbra, Coimbra, Portugal; [6] Independent Consultant, Coimbra, Portugal

1.   Lessons from real-life implementation

This is not a book about good intentions and noble objectives. It is about the challenges of translating them into action. In the previous 11 chapters, we have seen examples of how difficult it is to implement disaster risk reduction (DRR)—even when stakeholders act in good faith. We have seen that the challenges of implementation are often enormous: People disagree on both objectives and methods to achieve them; governance structures are fragile; there are insufficient knowledge, information, and expertise; actions depend on political will; there is not enough capacity to navigate implementation barriers; and funding for disaster prevention is elusive.

The adverse social and economic effects of the long-lasting COVID-19 pandemic, with the significant reduction of funds for international aid and the concentration of human resources on the struggle against the biological disaster, illustrate these challenges. They also remind us of the complexity increasingly generated by the cascading nature of current crises. Yet, we know that preventing disruptive events in the context of frequent and concurrent disasters and increasing climate-change pressures requires understanding the root causes of disasters.

Allan Lavell reminded us in the Foreword that disaster prevention and preparedness have been on the global development agenda since the early works of UNDRO, the then UN agency in charge of disaster-relief operations during the 1970s and 1980s. In 1994, these notions were included in the World Conference on Natural Disaster Reduction (Yokohama, Japan). In the following decades, DRR approaches kept changing. The principle of investing right now to better deal with future hazards and risks became the core of disaster prevention in the 2005 Hyogo Framework for Action. The 2015 Sendai Framework marked a shift toward managing and preventing new risks that lead to disasters, in contrast with the previous focus on identifying risks and managing disasters' impacts (Izumi, 2017; Mizutori, 2020). This reorientation is welcome. However, the Sendai Framework still fails to propose feasible means to change deeply rooted economic-growth models and forms of development that create or perpetuate vulnerabilities. Our argument in this book has been that without clearly dealing with underlying sociopolitical risk drivers, the Sendai Framework is unlikely to succeed.

There is a consensus that investment in DRR requires collaborative work between civil society and the private and public sectors (Choudhary & Neeli, 2018; Mizutori, 2020; Wisner, 2020). Uncertainty about the occurrence of disasters, however, still hinders investments in preparedness, leading to an emphasis on financing emergency-response efforts after disasters (Izumi, 2017; Kunz et al., 2014; Muir-Wood, 2018). Five years after the adoption of the Sendai Framework, Mami Mizutori (2020), head of the UNDRR, acknowledged that although the importance of investing in disaster planning and preparedness is now universally accepted, several challenges still exist in global financing models. These include the lack of reliable information on how to fund DRR, appropriate policies, and political will. In addition, local interpretations and adaptations of the Sendai Framework are still needed to translate general objectives into concrete, operational, and localized outcomes (Faivre et al., 2018; Moure et al., 2021; Van Niekerk et al., 2020).

The cases presented in this book illustrated the diversity that exists in DRR initiatives and approaches. They highlighted in different ways common difficulties in the implementation of the Sendai Framework's Priority 3, as well as the gaps that still exist between academics, citizens, and politicians.

We have attempted to unpack the main challenges—but also opportunities—in the implementation of the Priority 3. We have done so by exploring the three broad domains under which the chapters were organized: architectural and urban design for DRR, new methods for resilience, and building DRR knowledge. Inclusiveness and contextualization in DRR are transversal ideas that permeate all the previous chapters. Here we present some of the key patterns that we found and we propose some recommendations.

2. The roles and challenges of design processes in disaster risk reduction

The first part of this book focused on architecture and urban-design processes in DRR and resilience investment. In three chapters, we tackled different dimensions of design and DRR. But the cases also shared a common trait: They exemplified how architecture and urban-design initiatives embedded in a humanitarian and/or risk approach constitute significant learning processes for the involved participants. By actively engaging in such design-led processes, members of vulnerable communities, technocrats, consultants, and heritage professionals alike can become DRR champions and make more informed decisions.

In Chapter 1, Lusterio, Matabang, and Tan Singco highlighted the importance of investing in design-oriented disaster prevention and capacity building for poor and vulnerable communities. The authors drew attention to community participation in architectural design, planning, and building in recovery contexts. They claimed that by working in consortiums, civil society organizations are better positioned to grasp grassroots potentials and needs, and thus to ensure more effective participation and make more efficient DRR investments. The authors questioned reactive and centralized types of DRR management, which are incompatible with communities' daily struggles. This problem is especially relevant in the Philippines and other countries of the Global South that are exposed to multiple hazards and have poor capacities to invest in disaster preparedness.

But blindly focusing on preparedness might not be enough. In Chapter 11, Surtiari, Garschagen, Mendes, and Budiyono presented an Indonesian counterexample that involves DRR interventions before disasters. Yet, they warned that taking a proactive DRR approach without considering the harsh living conditions of the urban poor can be as perverse and inefficient as reactive DRR investments. Although not comparable, the experiences presented in Chapters 1 and 11 highlighted how entangled DRR and development are in terms of decision-making and investments.

Another key theme that emerged in Chapter 1 is the localization of DRR approaches, a topic previously explored by several academics (Aronsson-Storrier, 2020; Lucci, 2015; Pearson & Pelling, 2015; Van Niekerk et al., 2020). But here we contend that it is vital to consider not only expected outcomes but also undesirable shortcomings. For instance, Sandholz et al. (2020) highlighted that even though local autonomy is welcomed, in the absence of integrated policy and planning, it can endanger coherence and alignment at the urban, regional, or national levels. Gaps or misalignments between local and national policies and agendas may even create new risks or exacerbate existing vulnerabilities (Aronsson-Storrier, 2020). The conclusions of Chapter 1 pinpointed that a more effective Philippine DRR system would probably be a blend of community-based initiatives (largely focused on mitigation and adaptation) and governmental ones (possibly related to preparedness and relief). Government action should aim at reversing the inequalities that feed vulnerability in the country. A similar argument was also put forward in Chapter 6, by Adekola and Lamond, and in Chapter 4, by Gotangco and Josol. Although the Sendai Framework recognizes the links between DRR, sustainable development, and poverty eradication, these can barely be managed without bringing risks' political dimensions to the forefront.

Architectural and urban design is not only a creative process in professional work. In DRR and climate action, architectural and urban design can be a learning process for a multiplicity of stakeholders, whereby goals are identified, discussed, and agreed upon. Two cases in which heritage is at stake—presented in Chapters 2 and 3—exemplified the importance of culture in DRR. Cultural resilience has been responsible for keeping alive not only long-lasting DRR practices (as shown in Chapter 2) but also heritage assets (as exemplified in Chapter 3). Both chapters illustrated how design can foster disaster resilience by incorporating potential disruptions as a central dimension to foster synergies between history and DRR while maintaining the touristic value of heritage sites.

In Chapter 2, Kano, Tanaka, and Gota explained how they applied the lessons learned in several Asian heritage towns in the elaboration of a disaster risk mitigation plan for Bergama (Turkey). Their plan emphasized the gathering role of social places in historic towns while paying attention to the needs of citizens, migrants, and visitors. In Chapter 3, Hobeica and Hobeica reminded us that DRR interventions in historic sites should properly consider contingencies without underrating the heritage's potential to build cultural resilience. Heritage buildings and surrounding public spaces are living legacies whose DRR roles can be better channeled through architectural and urban design. The authors of Chapters 2 and 3 highlighted that heritage places are at the same time fragile and sources of resilience building. Therefore, DRR investments should not only protect them but also harness their intrinsic protective role.

The cases presented in Chapters 1 and 2 stressed the need to put citizens at the core of DRR investments while understanding their local conditions and linking their objectives to regional and national agendas. The increasing complexity of, and interactions between, development and environmental goals suggest a fundamental reevaluation of how we invest in DRR. Such reevaluation must consider community engagement, but also action at the regional and national levels. If properly carried out, architecture and urban design can be not just an eye-catching result but also a process to align agendas at various scales, discuss differences, identify alternatives, and reach consensus toward common goals. The cases presented in this book suggested that they are key tools to fulfill the Sendai Framework's principle of "all-of-society engagement and partnership" (UNISDR, 2015).

3. Inclusive methods in disaster risk reduction

The second part of the book focused on investing in new, integrated methods for building resilience and achieving DRR. The methods illustrated in Chapters 4 and 7 acknowledged the complexities and feedback loops that exist in complex DRR processes. The cases shown in Chapters 5 and 6 stressed the difficulty of reaching consensus among stakeholders about DRR goals and means to achieve them. Unlike what many international frameworks intend us to believe, DRR principles and priorities are different among citizens, politicians, and academics, and so are the interpretations and perceptions of risks, disasters, and climate-change impacts within and across communities. Such differences can hinder communication and the pursuit of shared aspirations. Methods that mitigate these barriers and promote horizontal dialog among citizens, academics, and decision-makers are hence welcome.

Gotangco and Josol, in Chapter 4, presented a holistic method for flood-hazard assessment, which they illustrated with an application in Manila, the Philippines. By considering multiple urban and human pressures causing disastrous floodings, they addressed the systemic nature of risk drivers. Furthermore, the application of their Physical Services Index revealed the importance of linking development approaches and climate change to avoid the creation of new risks. Dealing with the same geographic and institutional contexts, but taking very different research standpoints, the authors of Chapters 4 and 1 reached similar conclusions. They emphasized that in times of climate change, when hazards may follow unexpected pathways and tend to create cascading effects, a DRR stance that focuses on "people first"— that is, on vulnerability reduction—is necessary. This reasoning is aligned with Kelman's standpoint that disaster prevention "should positively and tangibly impact day-to-day living" (Kelman, 2019, p. 2). Thus, governments and citizens should not see DRR exclusively as a means to face extraordinary situations.

Focused, likewise, on flood risk, Cardoso, Almeida, Telhado, Morais, and Brito presented, in Chapter 7, an objective-driven framework for assessing resilience. Taking Lisbon's water and waste services as examples, the authors examined how urban systems can become increasingly involved in cascading effects in times of climate change. They showed the importance of systematic analyses of hazard and risk mapping, enabling the development of resilience strategies for quick recovery and opportunities for appropriate DRR investments. In this way, the study evidenced the effectiveness of incorporating interdependencies between diverse urban systems. Their Resilience Assessment Framework can be a tool to localize the Sendai Framework at the city level while providing coherence with other post-2015 agendas to reduce urban risks.

Localizing and adapting the Sendai Framework to specific conditions at the urban and city levels have been the subject of multiple efforts since 2015 (UNDP, 2015). The Sendai Framework encourages countries and regions to better understand the multiple risks they are exposed to (Mysiak et al., 2018). Localizing the Sendai Framework, some experts believe, can also lead to a more effective evaluation of the local progress at regional, national, and subnational levels (Lucci, 2015). In fact, many authors argue that the Sendai Framework presents opportunities for framing national, regional, and local policies on DRR (Pearson & Pelling, 2015). For some of them, by localizing the Sendai Framework to address disaster risks, policy- and decision-makers can also address key risk drivers such as inequalities and poverty (Lucci, 2015). Cardoso and colleagues, in Chapter 7, illustrated the influence that the European Union's priorities and principles have had on implementing climate-action projects at the local and inter-organizational levels. They showed how political influence on DRR investments reverberates at different scales—geographically, ideologically, and technically—and concluded that it is possible to have a multinational approach that is properly applied locally.

Political receptivity or resistance plays an essential role in the implementation of DRR initiatives. Political and ideological influences can profoundly affect local communities, as Buckman explained in Chapter 5. His experience in working with antagonistic and conservative communities in the United States evidenced the shortcomings that planners face when transitioning from policy to practice. Buckman found that it is sometimes necessary to rethink DRR strategies and investments and to use innovative tools to negotiate with communities. Reaching consensus when diverging standpoints prevail requires tackling communities' prime concerns such as livelihoods. This chapter also stressed the weight of financial matters in DRR decisions in liberal contexts such as the United States, an issue that somehow sheds light on the Sendai Framework's emphasis on the DRR's financial approach. Without such emphasis, governments, institutions, and citizens might not endorse reducing disaster risks or see it as a priority. DRR cannot become only a matter of protecting business interests or adjusting socioenvironmental systems through financial and technological means. It requires a critical position about what those systems ought to be and whether and how they are accessible to everybody. It is therefore necessary to consider structural factors behind risk-creation processes, notably social and environmental injustices, and to revise our responses to climate change adaptation (CCA).

Adekola and Lamond did exactly that in Chapter 6. Challenging the conventional CCA engagement practices, they suggested a systems-thinking communication framework that invites stakeholders to embrace change instead of avoiding it. They argued that stakeholders' engagement in CCA provides the opportunity to better understand the complexities of climate-related risks and impacts. Conceived as an empowerment means, the framework proposed in Chapter 6 entails an engagement process in which multiple parties can interact in horizontal governance structures. By defining specific steps—from the identification and gathering of concerned people to communication—the authors described a process that somehow resembles the Community Action Teams presented in Chapter 5.

Chapter 6 also reminded us that the lack of coherence among agendas pursued by policy-makers and agencies leads to fierce competition for implementation funding. Poor articulation between agendas often leads to excessive focus on one or a few objectives and implementation delays in others (Wisner, 2020). There are also duplication of intervention costs that ultimately undermine overall DRR (Moure et al., 2021). It is therefore vital to harness comprehensive implementation and to enable synergies and collaborative actions to avoid wasting scarce resources. In many cases, it is crucial to recognize and manage trade-offs between different goals and to try to maximize gains in an ethical manner that respects the needs and aspirations of the most vulnerable.

Another key idea underlying Chapter 6 is that investing in DRR and climate-change resilience often implies massive monetary outlays, a process that is thus potentially biased to meet the interests of the wealthy. Investors and sponsors may perceive DRR as any other investment: a source of financial profit. The needs and expectations of marginalized citizens, minorities, poor communities, and other deprived groups tend to be sidelined by economic elites. It is therefore crucial that they be supported by government-led initiatives that favor collective goals instead of private privileges.

Overall, the chapters in Part B supported the idea that instead of simplifying complex issues to turn them tractable, stakeholders must recognize intricacies and contingencies as a precondition to pursuing effective and durable DRR. An overarching argument throughout this book was that consensus about DRR goals and means rarely exists among heterogeneous stakeholders. Yet, the cases presented in this part suggested that inclusive methods based on engaged dialog and a deep understanding of, and interest in, particular contextual conditions are not only possible but necessary. More inclusive DRR practices are required to avoid siloed problem framings, biased attitudes, and investment decisions that only benefit the privileged.

4. Building and sharing knowledge on disaster risk reduction investment

The Sendai Framework was developed to consolidate and ensure continuity in the DRR policy formulated in the previous four decades. It fails to provide, however, specific details or principles about means of implementation. In this way, the Sendai Framework allows for multiple interpretations of targets and commitments. In many cases, this ambiguity hinders priorities and budgets at different administrative levels and creates misalignments between local priorities and national goals (Moure et al., 2021). Lack of clarity about methods and practices also undermines the capacity of stakeholders to implement the Sendai Framework in synergy with local conditions and regulations. Moreover, investments in DRR are rarely a linear process in which policy naturally leads to successful implementation, monitoring, and evaluation. Even well-elaborated DRR policies often find tortuous ways toward implementation and seldom meet their goals entirely. If we recognize that policy is not enough, building knowledge on the conditions that hinder DRR implementation is therefore key.

This is precisely the purpose of the third part of this book, focused on investing in knowledge as a key objective to enable evidence-based DRR decision-making. Egbelakin, Ogunmakinde, and Carrasco, in Chapter 8, analyzed the perspectives of heritage-building owners in New Zealand when it comes to retrofitting their properties. The study found that there were limited government-sponsored financing mechanisms to help retrofit and preserve heritage properties. By identifying gaps in the communication of existing incentives, the authors called for better risk-information sharing through feedback systems that capture the opinions of building owners. The authors concluded that effective DRR investments require awareness-raising and communication methods that allow for feedback loops while considering hindering factors such as different languages and values, as well as asymmetric power relationships between stakeholders. Like Chapter 5, Chapter 8 stressed the weight of financial drivers in individual DRR decisions among the well-off. The authors' results highlighted that ambiguities and deadlocks in implementation are not only a problem of countries in the Global South. DRR decision-making and investments in the Global North are likewise challenged by misaligned agendas, contradictions, and tensions.

The intricacies around DRR communication are also explored in Chapter 10, in which Tavares, Areia, Mendes, and Pinto analyzed how the Portuguese media produce and disseminate CCA information. Public awareness is a precondition to civil society's effective engagement and actions. But risk awareness is threatened when scientific information is dismissed in favor of policy-makers' interests. Under these circumstances, the Portuguese citizens might have fewer reasons to adhere to (or call for more effective) DRR and CCA initiatives, whose investments are more likely to fail than to succeed.

Sridarran, Keraminiyage, and Amaratunga, in Chapter 9, put in evidence the failures of resettlement management to thoroughly address the expectations of disaster-affected people in Sri Lanka. To circumvent resettlement inadequacies, the authors suggested a stronger involvement of the resettlers in the decisions related to their built environment. This is in line with the arguments by the authors of Chapter 1, who also dealt with postdisaster settings in the Global South. Both chapters emphasized how entangled DRR governance and legal systems are in contexts marked by higher vulnerabilities and fragile institutions.

Resettlement dissatisfaction often suggests that resources could have been invested elsewhere more efficiently (Lizarralde et al., 2010). Yet, as the pressures of climate change intensify, policy- and decision-makers in the Global South are increasingly compelled to consider relocation. In this context, they should not underestimate the negative consequences and secondary effects of resettlement programs. Instead, stakeholders must remember that relocation often engenders new risks and may increase the vulnerability of some individuals, businesses, and social groups, as exemplified in Chapter 11.

The Sendai Framework emphasizes the importance of implementing structural and nonstructural measures to enhance economic, social, health, and cultural resilience. But some governments still prioritize investments in massive urban upgrading, buildings' retrofitting, and protective infrastructures, such as seawalls and retaining basins (Izumi, 2017; Lizarralde, 2021). Structural investments are often costly and sometimes produce negative impacts on low-income and marginalized communities (Kunz et al., 2014). Chapter 11 exemplified the drawbacks of this hard-engineering mindset. Here the authors explored vulnerability drivers and the adaptive capacities of communities living in flood-prone areas in Jakarta, Indonesia. Surtiari, Garschagen, Mendes, and Budiyono observed that governmental measures to deal with, and adapt to, floods often undermine the efforts and capacities of residents living in informal settlements. Such measures might even end up increasing risks among marginalized and low-income communities in the long term. They can also accentuate environmental injustices by dismissing existing informal practices and structures.

Kunz et al. (2014) and Izumi (2017) have previously highlighted the impact on DRR of nonstructural (or intangible) measures, such as education, information capacity building, and communication. It is necessary to rethink the approaches for investing in DRR, balancing structural and nonstructural measures (Egbelakin et al., 2017). This can be achieved, for instance, by combining protective infrastructure or retrofitting projects with capacity building and information sharing, as exemplified in Chapter 1.

Throughout this book, we have seen that effective DRR is rarely attained through the construction of new infrastructures alone. To reduce risks, it is necessary to deal with their root causes, such as racism, colonialism, imperialism, elitism, tribalism, marginalization, exclusion, segregation, savage forms of capitalism, and other manifestations of social and environmental injustices. As illustrated in Chapters 1, 9, and 11, disregarding the inequalities that lead to the very creation of vulnerabilities, risks, and disasters hampers the sustainability of DRR investments. A framework that fails to deliver a clear message about addressing these injustices is likely to accomplish very little in the long run.

5. Disaster risk reduction is never easy, but investing in it is always crucial

Priority 3 of the Sendai Framework emphasizes the importance of developing sustainable financing mechanisms to enhance resilience. The COVID-19 pandemic has illustrated once again the cascading effects of crises, resulting in significant financial pressures, insufficient social and health services, and increases in inequalities and poverty. These problems are not merely consequences of the pandemic, but also part of the very causes of vulnerability that lead to crises and disasters. When evaluating investments in DRR, it is vital to consider the cascading impacts of hazards and the costs and effects of insufficient investment and inaction.

The cases presented in this book exemplified the real tensions and controversies that emerge in DRR investment. They also revealed the multiplicity of voices that exist and need to be heard. We conclude that many professionals and officers in government, international agencies, and NGOs still tend to neglect the complexity of these tensions and the multiplicity of interests and expectations at stake. Different local conditions require analyzing how stakeholders are involved, the specific needs and aspirations of affected citizens, as well as how to align DRR actions with local and national objectives.

Throughout this book, we have sustained that designing and implementing effective DRR interventions require contextualized solutions and tackling disasters' root causes and the cascading effects of risks. Only through a deep understanding of local built and socioeconomic environments it is possible to reduce vulnerabilities and enable meaningful actions to recover in the aftermath of a disaster.

It is fundamental to promote a context-oriented approach to DRR investment, international development, and humanitarian assistance, with special attention to issues in the Global South. International assistance for DRR must be based on the respect of local values, knowledge, capacities, and expectations. A focus on local leadership and expertise is needed to avoid biased investments and undesirable secondary effects. National governments need to develop risk-informed strategies to finance and plan DRR interventions, and to establish platforms for active multistakeholder involvement. Furthermore, government officers must acknowledge local stakeholders' leadership and the tense relationships that might exist between civil society and the state.

We argued in this book that it is crucial to shift our focus from disasters' consequences toward addressing their root causes. We are not alone. Previous studies have found, for example, that centralized decision-making, weak institutional structures and governance, political instability, and corruption have prevented effective DRR in Peru (French et al., 2020). Likewise, inappropriate forms of development and urbanization, as well as inefficient protective infrastructures, have misguided people's attitudes during disasters in Japan (Nakasu et al., 2018). It has become clear that disaster practitioners must still pay greater attention to local underlying risk drivers and the engagement of various relevant stakeholders.

6. The need to redress social and environmental injustices

Today, the Sendai Framework is part of a family of numerous concomitant global agreements. Investing in DRR is challenged by the heterogeneous character of goals set in such agreements and the consequent conflicts and competition for financial resources that exist today. We argue in this book that the Sendai Framework suffers from several problems previously found in international policy. In particular, the Sendai Framework

- focuses too much on good intentions while neglecting implementation challenges. But the devil often hides in the implementation details, not in the general slogans written in policy documents;
- assumes that consensus about goals and means exists. We know now that this consensus doesn't exist. Decision-makers must recognize these differences and the tensions they create;
- assumes that general principles are applicable at large. Yet, several principles written in international policy are not easily applicable on the ground; notably because interpretations, meanings, and representations of the same terms and concepts vary within and across communities;
- assumes that investment in DRR is a linear process—as if following objectives would inevitably produce the desired outcomes. But DRR processes are not linear: Tensions, conflicts, controversies, and changes complexify investment decisions and implementation, and create different paths for which there is no clear roadmap;
- fails to recognize the root causes of disasters. It fails to assume the political component of risks and therefore masks the factors that generate social and environmental injustices.

Understanding where and how to invest in DRR is crucial. However, the empirical evidence gathered in this book reaffirms that there are no universal answers and there is a wide range of investment possibilities in the name of resilience. The cases presented in this book reminded us that different understandings of risks, design, development, disasters, climate change, and daily threats hamper the implementation of the Sendai Framework's third priority. They also showed the major challenge that still exists in the transition from a global framework of good intentions to effective socially and culturally relevant local actions. These cases evidenced the need for inclusive design and building practices and investment in knowledge for informed decision-making, as well as systemic and integrated approaches for investing in DRR. So here is our main conclusion: Decision-makers truly engaged in investing in preventing disasters must inevitably tackle inequalities, racism, colonialism, imperialism, neoliberalism, elitism, tribalism, marginalization, exclusion, segregation, and other social and environmental injustices.

References

Aronsson-Storrier, M. (2020). Sendai five years on: Reflections on the role of international law in the creation and reduction of disaster risk. *International Journal of Disaster Risk Science, 11*(2), 230–238. https://doi.org/10.1007/s13753-020-00265-y

Choudhary, C., & Neeli, S. R. (2018). Risk governance for improving urban disaster risk reduction. In I. Pal, & R. Shaw (Eds.), *Disaster risk governance in India and cross cutting issues* (pp. 211–224). Singapore: Springer.

Egbelakin, T., Wilkinson, S., Ingham, J., Potangaroa, R., & Sajoudi, M. (2017). Incentives and motivators for improving building resilience to earthquake disaster. *Natural Hazards Review, 18*(4), 04017008. https://doi.org/10.1061/(ASCE)NH.1527-6996.0000249

Faivre, N., Sgobbi, A., Happaerts, S., Raynal, J., & Schmidt, L. (2018). Translating the Sendai Framework into action: The EU approach to ecosystem-based disaster risk reduction. *International Journal of Disaster Risk Reduction, 32*, 4–10. https://doi.org/10.1016/j.ijdrr.2017.12.015

French, A., Mechler, R., Arestegui, M., MacClune, K., & Cisneros, A. (2020). Root causes of recurrent catastrophe: The political ecology of El Niño-related disasters in Peru. *International Journal of Disaster Risk Reduction, 47*, 101539. https://doi.org/10.1016/j.ijdrr.2020.101539

Izumi, T. (2017). Investing in disaster risk reduction: Implications for science and technology based on case studies from the local and national governments, the private sector and a university network. In R. Shaw, K. Shiwaku, & T. Izumi (Eds.), *Science and technology in disaster risk reduction in Asia: Potentials and challenges* (pp. 223–237). London: Academic Press. https://doi.org/10.1016/B978-0-12-812711-7.00014-6

Kelman, I. (2019). Axioms and actions for preventing disasters. *Progress in Disaster Science, 2*. https://doi.org/10.1016/j.pdisas.2019.100008

Kunz, N., Reiner, G., & Gold, S. (2014). Investing in disaster management capabilities versus pre-positioning inventory: A new approach to disaster preparedness. *International Journal of Production Economics, 157*, 261–272. https://doi.org/10.1016/j.ijpe.2013.11.002

Lizarralde, G. (2021). *Unnatural disasters: Why most responses to risk and climate change fail but some succeed.* New York, NY: Columbia University Press.

Lizarralde, G., Johnson, C., & Davidson, C. (Eds.). (2010). *Rebuilding after disasters: From emergency to sustainability.* Oxford: Spon Press.

Lucci, P. (2015). *'Localising' the post-2015 agenda: What does it mean in practice?* London: Overseas Development Institute.

Mizutori, M. (2020). Reflections on the Sendai Framework for Disaster Risk Reduction: Five years since its adoption. *International Journal of Disaster Risk Science, 11*(2), 147–151. https://doi.org/10.1007/s13753-020-00261-2

Moure, M., Sandholz, S., Wannewitz, M., & Garschagen, M. (2021). No easy fixes: Government workers' perception of policy (in)coherence in the implementation of the post-2015 agenda in Mexico. *Climate Risk Management, 31*, 100270. https://doi.org/10.1016/j.crm.2020.100270

Muir-Wood, R. (2018). *Disaster risk auditing using probabilistic catastrophe loss modeling.* American Geophysical Union Fall Meeting 2018, abstract #PA22A-12. Retrieved from https://agu.confex.com/agu/fm18/meetingapp.cgi/Paper/391082.

Mysiak, J., Castellari, S., Kurnik, B., Swart, R., Pringle, P., Schwarze, R., ... van der Linden, P. (2018). Brief communication: Strengthening coherence between climate change adaptation and disaster risk reduction. *Natural Hazards and Earth System Sciences, 18*(11), 3137–3143. https://doi.org/10.5194/nhess-18-3137-2018

Nakasu, T., Ono, Y., & Pothisiri, W. (2018). Why did Rikuzentakata have a high death toll in the 2011 Great East Japan Earthquake and Tsunami disaster? Finding the devastating disaster's root causes. *International Journal of Disaster Risk Reduction, 27*, 21–36. https://doi.org/10.1016/j.ijdrr.2017.08.001

Pearson, L., & Pelling, M. (2015). The UN Sendai Framework for Disaster Risk Reduction 2015–2030: Negotiation process and prospects for science and practice. *Journal of Extreme Events, 2*(1), 1571001. https://doi.org/10.1142/S2345737615710013

Sandholz, S., Wannewitz, M., Moure, M., & Garschagen, M. (2020). *Costs and benefits of (in)coherence: Disaster risk reduction in the post-2015 agendas. Synthesis report.* Bonn: United Nations University Institute for Environment and Human Security.

UNDP (United Nations Development Programme). (2015). *The world we want: Consultations on the localization of the post-2015 development agenda.* Concept paper. New York, NY: UNDP.

UNISDR (United Nations Office for Disaster Risk Reduction). (2015). *Sendai Framework for Disaster Risk Reduction 2015–2030.* Retrieved from https://www.preventionweb.net/files/43291_sendaiframeworkfordrren.pdf.

Van Niekerk, D., Coetzee, C., & Nemakonde, L. (2020). Implementing the Sendai Framework in Africa: Progress against the targets (2015–2018). *International Journal of Disaster Risk Science, 11*(2), 179–189. https://doi.org/10.1007/s13753-020-00266-x

Wisner, B. (2020). Five years beyond Sendai—Can we get beyond frameworks? *International Journal of Disaster Risk Science, 11*(2), 239–249. https://doi.org/10.1007/s13753-020-00263-0

Index

'*Note:* Page numbers followed by "f" indicate figures and "t" indicate tables.'

Printed in the United States
by Baker & Taylor Publisher Services